T0210719

MEDIA AND CLIMATE CHANGE

This book looks at the media's coverage of Climate Change and investigates its role in representing the complex realities of climate uncertainties and its effects on communities and the environment.

This book explores the socioeconomic and cultural understanding of climate issues and the influence of environment communication via the news and the public response to it. It also examines the position of the media as a facilitator between scientists, policy makers and the public. Drawing extensively from case studies, personal interviews, comparative analysis of international climate coverage and a close reading of newspaper reports and archives, the author studies the pattern and frequency of climate coverage in the Indian media and their outcomes. With a special focus on the Western Ghats, the book discusses the political rhetoric, policy parameters and events that trigger a debate about development over biodiversity crisis and environmental risks in India.

This book will be of great interest to scholars and researchers of environmental studies, especially Climate Change, media studies, public policy and South Asian studies, as well as conscientious citizens who deeply care for the environment.

Deepti Ganapathy is a faculty member at the Indian Institute of Management Bangalore and a former journalist who has widely covered and reported on infrastructure, health, women's issues, education and the environment. She holds a PhD in Communication and Journalism from the University of Mysore. She has received recognition for her work, including Journalist of the Month (February 2017) from the International Journalists' Network (ijnet), and was a finalist for the CNN Young Journalist Award (2007), among others. She was appointed Visiting Scholar at the University of California San Diego (2019) and was a Rotary GSE Fellow (2009) to New Zealand. She has presented her work at leading international conferences and given academic talks to government and business organizations. She has published over 1,000 articles across academic journals, book chapters and mainstream media. Her thought leadership articles and guest blogs continue to appear in popular newspapers and magazines, such as *Forbes*, *Deccan Herald*, *The Times of India*, *The Indian Express* and *The Economic Times*.

Deepti Ganapathy's writing is accessible – which makes this book relevant to politicians, journalists, academics, climate activists and even artists like me. It provides a balanced and informed analysis of India's environmental governance that focuses on managing the impacts of development while staring at a biodiversity crisis. With a much-needed positive tone and in a lucid manner, the reader gains a deep and rich insight on the conundrum that the media faces while mediating between the policy makers and environmentalists as they unravel the multidimensional and complex nature of climate change in India.

Ricky Kej, *Grammy Award Winner, UNCCD "Land Ambassador"*

In this engaging and comprehensive book, Deepti Ganapathy looks at climate change in the unique and fascinating locale – the Western Ghats, home to around 50 million people. This community is nestled in an ecosystem that is defined as one of the world's eight "hottest hotspots" of biological diversity. How does climate change impact this vulnerable part of the world, and how is it understood by its inhabitants? Focusing more on the latter question, Ganapathy explores press narratives and media reporting of climate change and how they impact public and political discourse.

The media act as a historical record relevant to its readership. Far from merely observer, the media is both an outcome and an antecedent of social debate. Recognizing this duality, it is vital to recognize, as Ganapathy does, that the media is an organizational actor whose output is subject to multiple political pressures. The book appropriately takes an expansive view of what we call "media" today, examining both traditional and digital media. And, in a very interesting twist, explores how the "digital divide" poses challenges for marginalized communities to have voice in the climate debate in India. And yet, as Ganapathy shows, these communities (specifically forest dwellers) have the closest connection to the environment and are "rich storehouses of knowledge that has been passed down from generations." In a unique and important facet of this book, Ganapathy explores how they construe and make sense of the news coverage related to climate change and how can they become effective activists in this narrative if the media would mine them as a credible source. Overall, this book can lead to collective Climate Action. I recommend this book for both scholars and practitioners alike.

Professor Andrew Hoffman,
Holcim (US) Professor of Sustainable Enterprise,
University of Michigan

The media has a critical role in engaging the public on the implications of climate change. This book plays an important role in setting out the evolving recognition of this role. Most people understand key issues based on what they see in the media; this places an important responsibility on journalists to be evidence based in their representations. This helps key policy initiatives gain public support. Protecting a global jewel like the Western Ghats is the prize if this is achieved, and that seems entirely worthwhile!

Dr. Marie Doole, *Environment policy researcher, New Zealand*

MEDIA AND CLIMATE CHANGE

Making Sense of Press Narratives

Deepti Ganapathy

Routledge
Taylor & Francis Group

LONDON AND NEW YORK

First published 2022
by Routledge
2 Park Square, Milton Park, Abingdon, Oxon OX14 4RN

and by Routledge
605 Third Avenue, New York, NY 10158

Routledge is an imprint of the Taylor & Francis Group, an informa business

British Library Cataloguing-in-Publication Data
A catalogue record for this book is available from the British Library.

Library of Congress Cataloging-in-Publication Data
A catalog record for this book has been requested.

ISBN: 978-0-367-44318-4 (hbk)
ISBN: 978-1-032-14925-7 (pbk)
ISBN: 978-1-003-01567-3 (ebk)

DOI: 10.4324/9781003015673

Typeset in Sabon
by Apex CoVantage, LLC

I dedicate this book to my parents who teach me each day to be caring, conscious and courageous – traits which they lead by example . . .

CONTENTS

FIGURES

TABLES

PREFACE

My interest in the field of media representations of Climate Change was kindled when I participated in a United Nations Foundation (UNF) initiative in 2016 to work with journalists around the globe toward the realization of the Sustainable Development Goals (SDGs).

After doing several programs with UNF, SDGs 13, 14 and 15 have time and again resonated in my mind for the urgency in the messages "Take urgent action to combat climate change and its impact," "Conserve and sustainably use the oceans," and "Protect, restore and promote sustainable use of terrestrial ecosystems, sustainably manage forests, combat desertification, and halt and reverse land degradation and halt biodiversity loss." Through these messages, which SDGs stress upon, there are a few overarching questions: Who speaks for the Silent Sentinels – we refer to as Nature? Who speaks for the climate? Who speaks for the forests? Who speaks for the ocean?

I attended an interesting Environmental Summit at the National Institute of Advanced Studies located at the Indian Institute of Science (IISc) campus in October 2017. The ideas and thoughts generated during this summit, which was co-organized by a Grammy winner – Ricky Kej – who has made it his mission to spread messages about conservation and environmental issues through his music, have been an eye-opener.

"For man cannot give wild animals freedom, he can only take it away," "Man's consumption patterns coupled with population explosion will consume the entire planet." "When you are on Mt. Everest or up in the North pole, you will realize the immense power of nature" – these were some of the stark comments from the speakers who were trying to burst the bubble that we live in.

We continue to complacently sit inside this bubble, waiting for nature to work its miraculous powers to restore and heal itself. When do we take charge? When should we get involved? What small baby steps can we take to save our planet? What is the legacy we want to leave behind for our future generations?

The initiative by UNF showed me the implications of the media's role in climate reporting. I committed to this UNF program to work on a news report and publish the report in my country with a leading media house. I found my sources and got the relevant data but was unable to find a leading media house to publish my story in their main pages.

The reality of not being able to communicate the most pressing issue of our times is disturbing. A more disturbing fact is that the implications of Climate Change are not reaching the most vulnerable sections of our society. The hands that grow our food are not empowered to understand the shifting dynamics of environmental issues. Our farmers and marginalized labor force do not understand that "no rain" or "more rain" is a cause of concern that could knockout their very survival.

When the capital of India is held to ransom by smog for a couple of days every year and media reporting focuses on blame game and sensationalization, it is a cause of concern. When news reports relating to Climate Change appear in the columns of main editions or are splashed on prime-time television news – if and only if there is a natural disaster, there is a political agenda or the news warrants coverage because world leaders are meeting in Paris or Bonn – it is worrisome.

The media's stance to cover news related to Climate Change has to shift gears from doomsday reporting to sustainable reporting. The media must feel a sense of their responsibility in empowering readers to make them feel a part of the problem and urge them to find a solution. This will lead toward "Climate Action." An empowered reader will feel encouraged to know the policies put forth by his government, and he will take active part in the process of policy formulation. An empowered citizenry will strength the process of policy making.

Governments have a major role to play in shaping the outcome of Climate Change, and they need to feel a sense of urgency, which can be brought out through consistent media reporting and people's intervention in the policy process.

I eventually found a place to publish my article – "Bees feel the sting of Climate Change" – not in the main pages, but in a supplement feature of a newspaper that has a weekly page dedicated to cover environment news. This was in 2017, and as I write this in 2021, the one-page feature in that newspaper has disappeared – due to lack of advertisements and readership.

This book is an opportunity to narrate through qualitative and quantitative work that I produced in 2017 on the intersections between media reporting of Climate Change and its implications on policy decisions. The major area of my work is situated in the Western Ghats. The Western Ghats can be considered as the microcosm of biodiversity. It is India's rainforests – a thriving ecosystem in the southwestern peninsula of India, rich in its biodiversity, a treasure house of endemic species, pristine and magnificent, spanning an area of 1,64,280 square kilometers.

Home to around 50 million people, the Western Ghats consist of 4,156 villages across six states, running north to south over a distance of 1,500 kilometers, with peaks ranging from 1,030 to 2,695 meters above sea level, rainfall ranging from 80 to 320 centimeters on an average. They are the source of major river systems, including the Cauvery, Krishna, Godavari, Palar and Pennar basins.

According to documented reports, the Western Ghats cover barely 5% of India's forest area but are home to 1,800 species endemic to the region and consists of 27% of all the species of higher plants recorded in the Indian region. They are also a hub of economic activity – where mini hydel power projects, tourism and many industries thrive.

The Western Ghats were declared a World Heritage Site by UNESCO in 2012. This tag brought along with it a myriad of restrictions on industries in the region. Why the Western Ghats are a macrocosm of Climate Change complexities is precisely what my research project unearths, as I delve into a five-year analysis of the coverage of media reporting on Climate Change in the period between 2012 and 2017.

Journalists, editors and media owners have a critical role to play in the narrative of Climate Change. The role of the media in addressing the challenges of global Climate Change should not be restricted to science or environmental coverage. This is an issue that is interdisciplinary at its core – affecting economies, communities and policy decisions. The role of the media must shift from disaster narratives to developing solutions.

I hope through this book, I can draw attention to the major themes and frames that emerge in media reporting. Journalists are faced with challenges from a range of actors – on the field, in the corridors of power and inside their establishments. Climate Change is a complex area of reporting, and unraveling the complexities of this issue is a daunting task that the media is facing. Today, Climate Change is not only an environmental issue, it has also become a people's issue and calls for action to be taken from ground-up.

The media must respond to this causality and tell the story of Climate Change in a manner that shakes the core of our very existence.

The idea for my article came from my larder – which showed depleting reserves of honey sourced from the Western Ghats. The fact that Climate Change was affecting my palate made me sit up and take notice. Bees are sensitive to variations in climate, and their extinction will have a chain effect on cross-pollination and food production – I learned. Therefore, it is important to establish the broader implications of Climate Change narratives to give the reader a "what's in it for me" and "how is this going to affect me," impression.

The media can demonstrate a key role in shaping the discourse of Climate Change, through communicating frequently upon public perception and policy agenda. This will influence the way in which Climate Change will be understood and acted upon by the public at large. In doing so, I have not been able to position this book based on international relations, trade and economics, as well as emerging and developed nations. This is where scholars who are working within their disciplines must engage in cross-national, interdisciplinary studies to unpack the layers of complexities that could lead to region-specific understanding of this problem so that policy interventions can be much more meaningful and specific.

The time is not too far away when future generations may never know the taste of honey, hear the croak of a frog, drink pure water from a mountain stream, smell the forest air or see a butterfly flitting across clear blue skies. They may be on their way to developing apps that give us a whiff of this sensory element! Through the insights that this book provides, I hope the media industry can invest in specialized coverage at the intersection of Climate Change, environmental policy and law to focus on covering the politics and policy of Climate Change, to help key decision-makers distill climate and environmental policy issues.

ACKNOWLEDGMENTS

My grandfathers – both maternal and paternal – shaped the dreamer and thinker in me. Kutta thatha – my maternal grandfather – built in me a curiosity for nature during my long summer vacations in Hanumantha estate, Kutta, Kodagu. A retired banker, he took to his coffee estate with a fierce professionalism. "Did you hear the call of the crow pheasant?" "A tiger has been here last night," "That is the karadi tree – the tallest and most majestic tree in my coffee estate," he would point out. I have preserved the many letters that my grandmother wrote to me, with sketches of the estate and the animals to keep me connected to nature. A part of me probably remains there, long after the family sold the estate. Recently, a neighbor of that estate who met me at a family wedding said, "Oh it's you Deepti! I can still picture your chubby baby face peering out of the gates of Hanumantha Estate when I drive by!" Mysore thatha – my paternal grandfather – an engineer with All India Radio who was one of the few in the world with a HAM radio, brought out the academic in me. He brought books two to three grades above – from CBSE, ICSE, Karnataka State curriculum – and made me solve mathematical equations, pushed me beyond my boundaries – wanting me to crack the competitive exams. With my parents' indulgence and their undoubted faith in my caliber, I grew up with a fierce competitive spirit but never let the dreamer in me idle. My brother, through his focus and discipline as a golfer and vice president of a company, taught me the value of diligence both on and off the field.

I started my career as a journalist with NDTV's Bengaluru bureau. One of my first trips as an intern with NDTV was to the iron ore mines in Hospet, Bellary, Karnataka. We were doing an expose on the use of child labor in the mines. Another editor from *The Hindu* joined us, she wrote a hard-hitting piece for the magazine *Frontline*, a sister concern of *The Hindu*, while we made several short and long-length feature stories for NDTV. In all this reporting the focus was on human rights violation and the nexus between politicians and profit-making businesses. What happened to the story of the environment? I wondered. The fingers of the women and men in the quarries were bleeding, but why couldn't the media reporters see beyond and speak for the rocks that were bleeding, for the trees that fell crying and for the rivers and water bodies into which rubble was thrown? After the exhausting two-day trip, we made a detour to see the ruins of Hampi, the erstwhile capital of the prosperous Vijayanagara Empire. I was lost in the magnificence and mystery among the ruins, drifting in my solitude from monument to monument, wondering about human greed. I was awakened from this dream when I heard our camera person call out my name. I hurried to the waiting car, only to be taken to task for wandering away. My mentor's concern for me, a 21-year-old wandering around in an isolated place, was genuine. How then can we expect young journalists to brave it all and enter unknown forests and habitations far away from the safety and comfort of their reporting desks and everyday buzz of political and crime news?

This book is an attempt to highlight the central role played by Press narratives not only in speaking for the environment and the nature of Climate Change but also in putting the spotlight on the crisis within newsrooms, of the editorial policies, such as the stance taken by the publication to choose to report or not to report on Climate Change. In doing so, I hope this book will expand the possibility of providing access to mainstream reporting of Climate Change and Climate Action to journalists wanting to do so and to showcase to the reader the spectrum of decision-making and implications that can come out of relentless reporting by the media – on the process of governance and policy decisions.

After a career of eight years with *Deccan Herald* and *The Times of India*, I took a new path as an academic. The first opportunity that I got for a research grant came from my alma mater. The University Grants Commission (UGC) had identified the University of Mysore as a University with Potential for Excellence (UPE), my proposal received funding through the Focus Area II of the UGC-UPE grant which was ably headed by Professor Usha Rani. I thank her for seeing in this topic a potential for greater understanding of why and how the reporting of Climate Change is produced, negotiated and disseminated in a highly imbalanced power nexus that exists in media systems in our country.

My mother, Indira Ganapathy, deserves recognition for ideating and accompanying me during the fieldwork expedition in 2017 into the Western Ghats. I thank her for always pushing me towards new horizons to discover my potential. I contacted many non-governmental organizations that work with tribal communities there, and I must thank the field officers for seamlessly arranging the logistics to ensure that after we had traveled deep into the settlements, we had a good number of respondents for our focus group discussions.

I thank my former employer – NMIMS University for extending on-duty leave to do my field work. Heartfelt gratitude to librarians Srinivas Murthy, Vasantha, Amit Joshi and Bheemashankar who not only helped me collect the archival data of newspapers but were a great source as local information providers, having lived and explored the length and breadth of the Western Ghats.

Finally, I hope that many young readers can emerge out of silos and get their hands and feet muddy with interdisciplinary work in this field, get into the forest, swamps, reefs, communities and mingle in these ecosystems to gain a true understanding of Climate Change and how nature communicates and communes naturally. This is nothing new, our ancestors had discovered the complex and intricate networks of this signaling mechanism. We do not have to reinvent the wheel. We merely have to scratch the surface, while unlearning and relearning all that we have come holding into the twenty-first century.

I dedicate this book to the innumerable, faceless and nameless reporters who have the courage and fortitude to pursue difficult choices and have the tenacity to question their editors when their stories get shelved. Rohith B.R. was one such reporter who made it his mission to chase scientists, environmentalists, forest officials and ordinary citizens to make sense of the narrative and published impactful articles in *The Times of India*. He passed away in an accident in February 2020. Environment-related reporting appears to have lost its spark along with him – the butterflies and frogs that he gave voice to, seem to be darting around in circles as their habitat gets stripped away.

1

WHY COMMUNICATE CLIMATE CHANGE?

Defining symbols and our relationship with the environment

"Earth provides enough to satisfy every man's need, not every man's greed."

~ *Mahatma Gandhi*[1]

The Polar bear, breathes in from the north, from the land of the frozen seas . . .
With its fur so white and its eyes so bright, now what can the secret be?

A picture of the polar bear standing on the last sheet of ice in the distant North Pole along with the headline "Be Worried, Be Very Worried" splashed on the 3 April 2006 cover of *TIME* magazine. The cover pages of the magazines *National Geographic* and *Wildlife* also showcased a picture of the polar bear with the headlines "Polar Vortex," "Global Warming," "Arctic Meltdown."

Climate Change scientists and scholarship that disregarded the notion that Climate Change is for real 20 years ago have today openly and publicly accepted that they were wrong and have plunged back into projecting the trends and areas that will shape the discourse in the coming years. The pandemic that we lived through in 2020 and continue to face in 2021 has become the fierce wind that gives strength to the glowing embers of a Climate Change narrative that was dying away. Today, world leaders – both political and business – are having conversations around the phrases "race to zero emissions," "reaching tipping point" and "human-made Climate Change."

The polar bear, in all its pure white and noble appearance, was appealing to the inhabitants of the entire planet, standing forlornly with its habitat endangered and no other place to migrate to. The subtle messages screamed loud through this frame. Around this frame the impact of Climate Change was meant to be conveyed to the world. The critical role of the centrality that mainstream media or newspapers and magazines played in the understanding of Climate Change can be interpreted in three distinct layers – first,

DOI: 10.4324/9781003015673-1

it was meant to highlight the role played by governments in combating the problem collectively; second, it directed attention toward susceptible regions, flora and fauna, as well as indigenous communities that were most affected by the ravages of Climate Change; and, third, it played a significant role in the projection of people who were emerging as green warriors championing for the cause of the environment through their activism and solutions that they offered on the field.

Here and throughout this book I focus on mainstream media – mainly the print medium – and situate my findings through the narratives of the Press.

Scholarship on news coverage of Climate Change has focused largely on the United States and Western Europe. Climate skepticism, which is predominantly the core of the narrative in these countries, is not necessarily addressed in developing economies. In Asia, the continent which will face the brunt of Climate Change due to its population density, the narrative has been restricted to coverage during calamities or major international environment summits. This book will not only draw attention toward current research in Climate Change coverage by the media but also critically examine the role played by the media to shape the discourse for government and non-governmental organizations to set the agenda. The book will appeal to the media and Climate Change scholars, journalists, policy makers and scientists to stop being mere spectators in this discourse but become active producers in the creation of news. This could be achieved by highlighting the significant role of Climate Action warriors, thereby shaping media coverage toward more proactive narratives rather than "doomsday" voices.

The purpose of writing this book is first to draw attention to the fact that reporting of Climate Change is not only a complex issue but a significant one. Media coverage of Climate Change issues, especially in Press narratives, can have direct policy implications. The coverage of protests and activism can directly or indirectly show how political elites, policy makers and local administration evaluate these protests as success stories. This voice of the public can serve as an opinion when it reaches government agencies through the channels of mainstream media and news media.

Second, since media coverage of Climate Change has implications for policy matters, it is critical to examine who is benefitting from this narrative and who is losing out. Here, the onus is on media houses and journalists themselves to produce and disseminate facts to the key stakeholders in a balanced and fair manner. Finally, it is interesting to analyze how Climate Change communication is dissected and decoded among the different stakeholders (scientists, communities, media and policy makers) and how they engage in this discourse.

An assessment of the 2012–2017 media coverage of the Western Ghats as a macrocosm of Climate Change complexities serves as an eye-opener to make sense of the vast and rich biodiversity that exists in India and the country's resurgence to improve its economy after freeing itself from colonial rule in 1947.

Asian countries are facing the brunt of Climate Change as they toe the line between the path of development and growth while adhering to COP protocols. The underlying concept of this book is to highlight the significant role played by the media in either underplaying or overly highlighting the communication of Climate Change.

This book relies on a research project that examined media representations of Climate Change through the coverage of the declaration of the Western Ghats as one of the world's biodiversity hotspots by UNESCO. The findings provide evidence that media coverage of issues related to Climate Change peaks during times of crisis. Media coverage was also found to be linked to the controversy surrounding the zoning of Ecologically Sensitive Areas (ESAs) in the Karnataka region of the Western Ghats. Furthermore, it was also concluded that media coverage peaks when issues related to overlapping themes of sustainability, policy decisions and Climate Change occur.

As we enter a new decade – a decade that begins with a pandemic that was never witnessed before – the discussions around Climate Change assume new meaning and significance. The interdependencies in nature, the symbiotic relationships and the need for interpreting and effectively communicating Climate Change to the various actors involved are calls for the media, governments and citizens to get involved in actively.

India's rainforests

Why should the Western Ghats, India's rainforests, which are a thriving ecosystem in the southwestern peninsula of India, rich in biodiversity, a treasure house of endemic species, pristine and magnificent, and spanning an area of 1,64,280 square kilometers be considered as a barometer for understanding the Press narratives around Climate Change?

Home to around 50 million people, the Western Ghats consist of 4,156 villages across six states, running north to south over a distance of 1,500 kilometers, with peaks ranging from 1,030 to 2,695 meters above sea level and rainfall ranging from 80 to 320 centimeters on an average. Furthermore, the Western Ghats are the source of major river systems, including the Cauvery, Krishna, Godavari, Palar and Pennar basins.

According to documented reports, the Western Ghats cover barely 5% of India's forest area but are home to 1,800 species endemic to the region and consist of 27% of all the species of higher plants recorded in the Indian region. A significant number of industries and resources are found along the Western Ghats, which traverses five states on the western coast – Maharashtra, Goa, Karnataka, Kerala and Tamil Nadu.

The Western Ghats were declared a World Heritage Site by UNESCO in 2012. The government constituted two expert panel committees to give recommendations on the scope and area of development after the ESAs were earmarked following the tag given by UNESCO. This resulted in political unrest and protests since crores of rupees were at stake for projects that would have to be shelved. Indigenous communities living inside the forests too were asked to evict, and their livelihoods were on the verge of being wiped away as these forests were now protected reserves. Over a period of five years, between 2012 and 2017, Press narratives display a range of themes and topics, giving ample scope for an analysis of the coverage of media reporting on Climate Change.

The story of Climate Change is a global one. Through the perspectives that journalism reflects in this five-year coverage, there are many lessons that are global in nature. Through the reporting, priorities and interests of key readers or audiences are often reflected. Often, the newspaper or media channel which has a wider reach also serves those vested interests. Climate Change affects people in the developing countries, and they are struggling to respond. The government and not-for-profit agencies that work in this sector can provide solutions, but the message and language around the Press narratives too need to be powerful.

The construct of the polar bear seems far-fetched for people in developing and least developed countries – many of them highly vulnerable to the ongoing impacts of Climate Change, such as extreme heat waves, rainfall, tropical storms and change in weather patterns. Lack of reporting on localized data on how Climate Change has affected the local economy may need to be translated in a simpler form for consumption of news in local communities. If access to local data, experts who can substantiate and provide quotes and funding to cover Climate Change is available at the disposal of every journalist, the Press narratives can show a great amount of direction and resolve. A steady flow of information that is factual and credible through mainstream media, can communicate the gravity of the situation relating to Climate Change. This consistency in reporting can impact the manner in which local bodies, non-governmental agencies and government can make sense of the strong signals emitting from ground-up to make key policy decisions.

Press narratives have inspired a cultural shift with artists and musicians using Climate Change, the most pressing problem of this generation, to communicate through their works. N. S. Harsha, a Mysore-based artist, created a canvas which was up for auction at Christies in 2007–2008. It depicted hundreds of empty plastic chairs strewn at an election rally in India, scattered paper cups and all the garbage generated after the rally is over. Ricky Key, Grammy winner, uses music to collaborate with artists around the world to spread key messages about conservation and wildlife.

Climate Change stories from around the world

"Climate change is a notoriously difficult subject for journalists to report on, for editors to maintain interest in, and for audiences to grasp" (Painter, 2010, p. 3). Because journalists, like scientists, look for objectivity and fairness in reporting, an interesting correlation exists between the increasing number of peer-reviewed scientific papers that are being published and the increase in these research papers being cited as evidence in media reports by journalists. There are various databases to search through which list the thousands of academic papers published each year. Amidst options such as Google Scholar and Web of Science, Scopus, the world's largest abstract and citation database of peer-reviewed literature, was identified for this study ("Analysis: The Climate papers," 2016).

In Scopus, we searched for any academic paper with the phrase "Climate Change" or "global warming" in its title, abstract or keywords – to look at both the top papers and all papers far beyond the top 100. The search yielded a total of almost 120,000 papers as of the beginning of June 2016. Looking at the countries where these institutions reside, there is a prominent leaning toward Western countries in the northern hemisphere. The United States and the United Kingdom dominate, with almost three-quarters of the top 100 papers. The fourth place goes to the paper that was reported by more news outlets than any other – "Future temperature in southwest Asia projected to exceed a threshold for human adaptability" – in *Nature Climate Change*. This study projected that conditions in the Middle East could become so hot and humid by the end of the century that being outside for more than six hours would be intolerable for humans. And in the fifth place is the *Science* paper, "Possible artifacts of data biases in the recent global surface warming hiatus," which finds the much-discussed "slowdown" in warming on the Earth's surface may not exist after all. The study was picked up in 118 stories in 77 news outlets. Elsewhere, in the Top 10, the seventh place goes to another *Nature Climate Change* article "Reaching peak emissions," which was published during the climate summit in Paris in early December 2015. Using preliminary figures for 2015, researchers found that the rapid increase in global carbon dioxide (CO_2) emissions over the last decade seemed to have stalled. The break in the emissions trend was mainly down to a drop in coal use in China, the study said. The paper generated 81 news stories in 65 publications. Closely behind in the eighth place is the *Science* article "Planetary boundaries: Guiding human development on a changing planet." The research found that human activity has pushed the Earth into a critical mode. Four out of nine "planetary boundaries" have now been crossed, the researchers said, with biodiversity loss, fertilizer use, Climate Change and land use all now exceeding the point where the risk of sliding into a "much less hospitable" world becomes high.

The paper was reported by 43 news outlets in 55 stories, tweeted by 662 people and made it onto the Facebook walls of 93 users. The *Nature* article "Mapping tree density at a global scale" by lead author Dr. Thomas Crowther at Yale University was ranked the third. This paper contains the striking estimate that there are three trillion trees on the Earth – some eight times more than previously thought. But the news isn't as good as it sounds – deforestation means our forests are shrinking by around 10 billion trees every year. Published in September, the paper generated 89 news stories from 81 publications, tweets from 950 accounts and 30 Facebook wall posts, giving it an overall score of 1,578.

The highest scoring climate paper comes from the very beginning of 2015: published on 08 January – and with an Altimetric score of 2,061 – it is the *Nature* article "The geographical distribution of fossil fuels unused when limiting global warming to two degrees" by Dr. Christophe McGlade and Professor Paul Ekins of University College London. The paper suggests that in order to keep the global temperature rise to no more than 2°C above preindustrial levels, 80% of known coal reserves, 50% of gas reserves and 30% of oil reserves must remain unburned.

In the runners-up place is "Climate change in the Fertile Crescent and implications of the recent Syrian drought" by lead author Dr. Colin Kelley from the University of California, Santa Barbara. This paper was published in the Proceedings of the National Academy of Sciences of the United States of America (PNAS). The research suggests that the severe drought in the region since 2006 was a catalyst for the Syrian conflict and that Climate Change has made such droughts in the region more than twice as likely. The study received a lot of media attention when it was originally published in March and was frequently cited throughout the year in debates over potential climate-conflict links. Perhaps, most notably, in an interview with Sky News, Prince Charles highlighted Climate Change as one of the contributing factors to the Syrian conflict, which was widely reported by the national media with varying degrees of accuracy. As per available reports, about 38 journals, 127 conferences, 400 workshops are currently dedicated exclusively to Climate Change and about 129,056 articles are being published on the current trends in Climate Change.

Why bother about Press narratives of Climate Change

When the anthropogenic origins of the phenomenon called "Climate Change" began with the increase in research activity from environmental scientists in the 1970s, it raised doubts about the benefits of human activity for the planet. Economists and developmental scholars felt that the resources being spent to mitigate Climate Change might as well be used for other human-related activities such as providing employment, seeking to end poverty and malnutrition, handling pandemics and so on. This group of scientists and policy makers are of the opinion that if the planet is eventually going to burn itself out, why should we direct and divert resources toward saving it.

On the other hand there are environmentalists, forest communities, indigenous people and pockets of civil society that want to turn the clock back and revere and reverse this process. They claim to be the voice of the planet, urging governments and policy makers to come to terms with the apocalypse that awaits them. This group wants more resources allocated for renewable energy, conservation of forest land and protection of indigenous territories and calls for global efforts toward Climate Action.

The story of the media representations of Climate Change begins with the picture of the lonely, forlorn-looking polar bear clutching the last piece of floating ice with a range of headlines that screamed aloud in bold – "Be Worried, be very worried," "Arctic Meltdown," "Polar Vortex." These images remained a powerful symbol of communicating Climate Change through the media. Three distinct messages were emerging out of this discourse 20 years ago:

1 Identity of the Earth at stake.
  Climate Change was occurring right here on planet Earth and was not a phenomenon occurring so far away in another distant galaxy, millions of light years away from us.
2 The endangered species becoming extinct.
  The implication of our actions and the consequences of our deeds were being felt at the remotest corners of the Earth by species that were most susceptible and fragile.
3 Doomsday scenario.
  The doomsday narrative resulted in two bifurcations of the Climate Change debate – one of denial and the other toward action.

Box 1 A brief history of Climate Change

1712 – British ironmonger Thomas Newcomen invents the first widely used steam engine, paving the way for the Industrial Revolution and industrial scale use of coal.

1800 – World population reaches one billion.

1824 – French physicist Joseph Fourier describes the Earth's natural "greenhouse effect." He writes: "The temperature [of the Earth] can be augmented by the interposition of the atmosphere, because heat in the state of light finds less resistance in penetrating the air, than in re-passing into the air when converted into non-luminous heat."

1861 – Irish physicist John Tyndall shows that water vapor and certain other gases create the greenhouse effect. "This aqueous vapour is a blanket more necessary to the vegetable life of England than clothing is to man," he concludes. More than a century later, he is

---

7

honored by having a prominent UK climate research organization – the Tyndall Centre – named after him.

1886 – Karl Benz unveils the Motorwagen, often regarded as the first true automobile.

1896 – Swedish chemist Svante Arrhenius concludes that industrial-age coal burning will enhance the natural greenhouse effect. He suggests this might be beneficial for future generations. His conclusions on the likely size of the "man-made greenhouse" are in the same ballpark – a few degrees Celsius for a doubling of CO_2 – as modern-day climate models.

1900 – Another Swede, Knut Angstrom, discovers that even at the tiny concentrations found in the atmosphere, CO_2 strongly absorbs parts of the infrared spectrum. Although he does not realize the significance, Angstrom has shown that a trace gas can produce greenhouse warming.

1927 – Carbon emissions from fossil fuel burning and industry reach one billion tonnes per year.

1930 – Human population reaches two billion.

1938 – Using records from 147 weather stations around the world, British engineer Guy Callendar shows that temperatures had risen over the previous century. He also shows that CO_2 concentrations had increased over the same period and suggests this caused the warming. The "Callendar effect" is widely dismissed by meteorologists.

1955 – Using a new generation of equipment including early computers, US researcher Gilbert Plass analyzes in detail the infrared absorption of various gases. He concludes that doubling CO_2 concentrations would increase the temperature by 3–4°C.

1957 – US oceanographer Roger Revelle and chemist Hans Suess show that seawater will not absorb all the additional CO_2 entering the atmosphere, as many had assumed. Revelle writes: "Human beings are now carrying out a large-scale geophysical experiment."

1958 – Using equipment he had developed, Charles David (Dave) Keeling begins systematic measurements of atmospheric CO_2 at Mauna Loa in Hawaii and in Antarctica. Within four years, the project – which continues today – provides the first unequivocal proof that CO_2 concentrations are rising.

1960 – Human population reaches three billion.

1965 –The US president's advisory committee panel warns that the greenhouse effect is a matter of "real concern."

1972 – The first UN environment conference was held in Stockholm. Climate Change hardly registers on the agenda, which center on issues such as chemical pollution, atomic bomb testing and whaling.

The United Nations Environment Programme (UNEP) is formed as a result.

1975 – Human population reaches four billion.

1975 – US scientist Wallace Broecker puts the term "global warming" into the public domain in the title of a scientific paper.

1987 – Human population reaches five billion.

1987 – Montreal Protocol agreed, restricting chemicals that damage the ozone layer. Although not established with Climate Change in mind, it has had a greater impact on greenhouse gas emissions than the Kyoto Protocol.

1988 – Intergovernmental Panel on Climate Change (IPCC) formed to collate and assess evidence on Climate Change.

1989 – UK Prime Minister Margaret Thatcher – possessor of a chemistry degree – warns in a speech to the UN that "We are seeing a vast increase in the amount of carbon dioxide reaching the atmosphere. . . . The result is that change in future is likely to be more fundamental and more widespread than anything we have known hitherto." She calls for a global treaty on Climate Change.

1989 – Carbon emissions from fossil fuel burning and industry reach six billion tonnes per year.

1990 – IPCC produces First Assessment Report. It concludes that temperatures have risen by 0.3–0.6°C over the last century, that humanity's emissions are adding to the atmosphere's natural complement of greenhouse gases and that the addition would be expected to result in warming.

1992 – At the Earth Summit in Rio de Janeiro, governments agree to the United Nations Framework Convention on Climate Change (UNFCCC). Its key objective was "stabilization of greenhouse gas concentrations in the atmosphere at a level that would prevent dangerous anthropogenic interference with the climate system." Developed countries agree to return their emissions to 1990 levels.

1995 – IPCC Second Assessment Report concludes that the balance of evidence suggests "a discernible human influence" on the Earth's climate. This has been called the first definitive statement to declare that humans are responsible for Climate Change.

1997 – Kyoto Protocol agreed. Developed nations pledge to reduce emissions by an average of 5% by the period 2008–2012, with wide variations on targets for individual countries. US Senate immediately declares it will not ratify the treaty.

1998 – Strong El Nino conditions combine with global warming to produce the warmest year on record. The average global temperature reached 0.52°C above the mean for the period 1961–1990 (a commonly used baseline).

1998 – Publication of the controversial "hockey stick" graph indicating that modern-day temperature rise in the northern hemisphere is unusual compared with the last 1,000 years. The work would later be the subject of two inquiries instigated by the US Congress.

1999 – Human population reaches six billion.

2001 – President George W. Bush removes the United States from the Kyoto process.

2001 – IPCC Third Assessment Report finds "new and stronger evidence" that humanity's emissions of greenhouse gases are the main cause of the warming seen in the second half of the twentieth century.

2005 – The Kyoto Protocol becomes international law for those countries still inside it.

2005 – UK Prime Minister Tony Blair selects Climate Change as a priority for his terms as chair of the G8 and president of the EU.

2006 – The Stern Review concludes that Climate Change could damage global GDP by up to 20% if left unchecked – but curbing it would cost about 1% of global GDP.

2006 – Carbon emissions from fossil fuel burning and industry reach eight billion tonnes per year.

2007 – The IPCC's Fourth Assessment Report concludes it is more than 90% likely that humanity's emissions of greenhouse gases are responsible for modern-day Climate Change.

2007 – The IPCC and former US Vice President Al Gore receive the Nobel Peace Prize "for their efforts to build up and disseminate greater knowledge about man-made climate change, and to lay the foundations for the measures that are needed to counteract such change."

2007 – At UN negotiations in Bali, governments agree to the two-year "Bali roadmap" aimed at hammering out a new global treaty by the end of 2009.

2008 – Half a century after beginning observations at Mauna Loa, the Keeling project shows that CO_2 concentrations have risen from 315 parts per million (ppm) in 1958 to 380 ppm in 2008.

2008 – Two months before taking office, incoming US President Barack Obama pledges to "engage vigorously" with the rest of the world on Climate Change.

2009 – China overtakes the United States as the world's biggest greenhouse gas emitter – although the United States remains well ahead on a per-capita basis.

2009 – Computer hackers download a huge tranche of emails from a server at the University of East Anglia's Climatic Research Unit and release some on the internet, leading to the "ClimateGate" affair.

2009 – A total of 192 governments convene for the UN climate summit in Copenhagen with high expectations of a new global agreement, but they leave with only a controversial political declaration, the Copenhagen Accord.

2010 – Developed countries begin contributing to a $30 billion, three-year deal on "Fast Start Finance" to help them "green" their economies and adapt to climate impacts.

2010 – A series of reviews into "ClimateGate" and the IPCC ask for more openness, but clear scientists of malpractice.

2010 – The UN summit in Mexico does not collapse, as had been feared, but ends with agreements on a number of issues.

2011 – A new analysis of the Earth's temperature record by scientists concerned over the "ClimateGate" allegations proves the planet's land surface has warmed over the last century.

2011 – Human population reaches seven billion.

2011 – Data show concentrations of greenhouse gases are rising faster than in previous years.

2012 – The Arctic Sea ice reaches a minimum extent of 3.41 million square kilometers (1.32 million sq. mi.), a record for the lowest summer cover since satellite measurements began in 1979.

2013 – The Mauna Loa Observatory on Hawaii reports that the daily mean concentration of CO_2 in the atmosphere has surpassed 400 ppm for the first time since measurements began in 1958.

2013 – The first part of the IPCC's fifth assessment report says scientists are 95% certain that humans are the "dominant cause" of global warming since the 1950s.

2015 – The United Nations Climate Change Conference, COP 21 is held in Paris, where 133 countries pledged to reduce carbon emissions by 1–2 degrees.

2016 – The United Nations Climate Change Conference, COP 22 is held in Marrakech; 11 new countries ratified the Paris Agreement, making the total number of countries 144.

2017 – The United Nations Climate Change Conference, COP 23 is held in Bonn. It focused on the role of the indigenous people in Climate Change.

2018 – The United Nations Climate Change Conference, COP 24 is held in Katowice, Poland.

2019 – The United Nations Climate Change Conference, COP 25 is held in Madrid. It was the longest in history and ended in deadlock and disappointment over most of the contentious issues.

2021 – The United Nations Climate Change Conference, COP 26 to be held in Glasgow.

Source: BBC News environment correspondent Richard Black traces key milestones, scientific discoveries, technical innovations and political action. (*20 September 2013, Science & Environment, BBC*)

Note

1 Quoted in Weber, T. (2004). *Gandhi as disciple and mentor* (p. 227). Cambridge, UK: Cambridge University Press.

2

WHY IS THE MEDIA SHYING AWAY FROM COVERING CLIMATE CHANGE?

Frontline warriors

When the president of a small island nation in the Pacific Ocean made an urgent appeal to world leaders that his island is sinking due to rising seawater at a historic climate summit in New York in 2014, the world woke up to the reality of Climate Change. Three years later, President Anote Tong reiterated his appeal – "My people are the polar bears of the Pacific." His case has been made with science, with doomsday predictions, with emotional pleas. But so far none of his appeals has been enough to convince the world to save Kiribati, the 33 tiny islands in the Pacific Ocean so vulnerable to Climate Change that they could soon be nonexistent. So, Anote Tong, the president of the island nation, has taken to comparing himself to a polar bear. Somehow, he said, these animals elicit more of a response than do the 100,000 human population of Kiribati when talking about the impacts of Climate Change. Further, Tong stated he can see how the bears, like the people of Kiribati, face the threat of losing their homes very soon:

> I know that in the past there was a lot of focus on the polar bears. In my attempt to get attention on our own situation I draw a comparison that what happens to the polar bears will also be happening to us in our part of the world,

The UN's science panel recently asserted that it was "very likely" that the Arctic would witness an ice-free summer in 2050. Glaciers are also shrinking fast. Tong said:

> I saw for myself with my own eyes the huge sheets of ice from the glaciers and imagined that if that was to melt then obviously it's going to mean a lot more trouble for us and a lot of trouble for other people.

A huge ice shelf has since broken away from the Antarctic continent in 2017.

DOI: 10.4324/9781003015673-2

President Tong made his appeal again – this time from the seat of democracy, Bengaluru, the capital of Karnataka – home to one of the World's 206 Natural Heritage Sites, *the Western Ghats.*

On 6 October 2017, President Tong delivered his address to an audience of elected representatives of the state, at Grammy winner Ricky Kej's concert to raise environmental consciousness. His concerns were received with astonishment and disbelief.

On many fronts Climate Change and its impacts are occurring faster than expected. The irreversible trend or the "tipping point" is now well documented by scientists. Rising sea water levels and emissions of greenhouse gases – these are just the tip of the iceberg. Kiribati Island needs global temperatures to be contained at below 1.5°C to survive into the next century. Currently, the target of international negotiations is 2°C.

According to Pittock (2009):

> Climate is critical to the world as we know it. The landscape, and the plants and animals in it, are all determined to a large extent by climate acting over long intervals of time. Over geological time, climate has helped to shape mountains, build up the soil, determine the nature of the rivers, and build flood plains and deltas. At least until the advent of irrigation and industrialisation, climate determined food supplies and where human beings could live.
>
> (p. 2)

Why does it matter?

In the growing debate about whether Climate Change is human-induced or not, scientific consensus on anthropogenic Climate Change is growing (Anderegg et al., 2010; Brüggemann and Engesser, 2017; Cook et al., 2013; Oreskes, 2004). When the IPCC was formed in 1988, and an Indian, Dr. Rajendra K. Pachauri, associated with the IPCC was awarded the Nobel Prize, the Indian media woke up and began reporting on the severity of the problem.

From the Stockholm environment summit in 1972 to COP 23 at Bonn in 2017, what has changed in the last 45 years? In a foreword for a book, Dr. Rajendra K. Pachauri notes:

> In my view, which is shared by a growing body of concerned citizens worldwide, climate change is a challenge faced by the global community that will require unprecedented resolve and increasing ingenuity to tackle in the years ahead. Efforts to be made would need to be based on knowledge and informed assessment of the future.
>
> (Pittock, 2009)

Pittock (2009) argues:

> Scientists believe the rapid warming in the last several decades is due mostly to human-induced changes to the atmosphere, on top of some natural variations. Climate change induced by human activity may occur due to changes in the composition of the Earth's atmosphere from waste gases due to industry, farm animals and land clearing, or changes in the land surface reflectivity caused by land clearing, cropping and irrigation.
>
> (p. 7)

Where are its effects seen?

The world's first Climate Change refugees will be the 102,000 people of Kiribati Island. The three-time president of the island nation Anote Tong made a fervent appeal to world leaders in December 2015 during his speech at the UN COP 21 conference in Paris: "Give us a proposition that will guarantee that our people will remain above the water."

How do we know what we know about the climate? Climate Change induced by human activity may occur due to changes in the composition of the Earth's atmosphere from industrial waste gases, farm animals and land clearing, or changes in the land surface reflectivity caused by land clearing, cropping and irrigation.

Industrial waste gases include several, such as CO_2, methane (CH_4) and oxides of nitrogen, that can absorb heat radiation (long-wave or infrared radiation) from the Sun or the Earth. When warmed by the Sun or the Earth, they give off heat radiation both upward into space and downward to the Earth. These gases are called "greenhouse gases" and act like a thick blanket surrounding the Earth. In effect, the Earth's surface has to warm up to give off as much energy as heat radiation as is being absorbed from the incident sunlight (which includes visible, ultraviolet and infrared radiation). Soot particles from fires can also lead to local surface warming by absorbing sunlight, but reflective particles, such as those formed from sulfurous fumes (sulfate aerosols), can lead to local cooling by preventing sunlight from reaching the Earth's surface.

Natural greenhouse gases include carbon dioxide and methane and water vapor. These help to keep the Earth some 33°C warmer than if there were no greenhouse gases and clouds in the atmosphere. Human activities have increased the concentrations of several greenhouse gases in the atmosphere, leading to what is termed as the "enhanced greenhouse effect." These gases include CO_2, CH_4 and several other artificial chemicals. The Kyoto Protocol, set up to begin the task of reducing greenhouse gas emissions, includes a package or "basket" of six main gases to be regulated. Besides CO_2 and

CH_4, these are nitrous oxide (N_2O), hydrofluorocarbons (HFCs), perfluorocarbons (PFCs) and sulfur hexafluoride (SF6).

Anthropogenic or human-caused increases in CO_2 come mainly from the burning of fossil fuels such as coal, oil and natural gas, the destruction of forests and carbon-rich soil, and the manufacture of cement from limestone. The concentration of CO_2 before major land clearing and industrialization in the eighteenth century was about 265 ppm. CH_4 comes from decaying vegetable matter in rice paddies, digestive processes in sheep and cattle, burning and decay of biological matter and fossil fuel production. HFCs are manufactured gases once widely used in refrigerants and other industries but which are largely being phased out of use because of their potential to destroy the atmospheric ozone. PFCs and SF6 are industrial gases used in the electronic and electrical industries, firefighting, and solvents and other industries.

The amplification comes from warmer oceans giving off dissolved CO_2, and thus increasing the natural warming via the greenhouse effect. As early as the nineteenth century, some scientists noted that increased emissions of CO_2 might lead to global warming.

Current estimates of future Climate Change are based on projections of future emissions of greenhouse gases and resulting concentrations of these gases in the atmosphere. These estimates also depend on factors such as the sensitivity of global climate to increases in greenhouse gas concentrations; the simultaneous warming or cooling effects of natural climate fluctuations; and changes in dust and other particles in the atmosphere from volcanoes, dust storms and industry.

Given that climate has changed during the twentieth century, the key question is how much of this Climate Change is due to human-induced increased greenhouse gas emissions and other more natural causes. This has great relevance to policy because, if the changes are due to human activity, they are likely to continue and even accelerate unless we change human behavior and reduce our emissions of greenhouse gases (Pittock, 2009).

A paper by William Ruddiman of the University of Virginia, in 2003, raises the possibility that human influence on the climate has been significant since well before the Industrial Revolution due to the cutting down of primeval forests to make way for agriculture and irrigated rice farming in Asia. Ruddiman claims that the Earth's orbital changes had led to a decline in CO_2 and CH_4 concentrations in the atmosphere 8000 years ago. Instead, there was a rise of 100 parts per billion in CH_4 concentrations and of 20–25 ppm in CO_2 by the start of the Industrial Revolution. He calculates that this has led to the Earth being 0.8°C warmer than if humans had not been active, an effect hidden because it has canceled out natural cooling due to orbital variations. Simulations of the response to natural forcing alone (i.e., natural changes causing the climate to change), such as variability in energy from the Sun and the effects of volcanic dust, do not explain the

warming experienced in the second half of the twentieth century. However, they may have contributed to the warming observed in the previous
50 years. The sulfate aerosol effect would have caused cooling over the last
half century, although by how much is uncertain. This cooling effect has
become less since the 1980s, as sulfur emissions have been reduced in North
America and Europe in order to control urban pollution and acid rain.

The best agreement between model simulations of climate and observations over the last 140 years has been found when all the aforementioned
human-induced and natural force factors are combined. These results show
that the factors included are sufficient to explain the observed changes, but
they do not exclude the possibility of other minor factors contributing. Furthermore, it is very likely that the twentieth-century warming has contributed significantly to the observed sea level rise of some 10–20 cm, through
the expansion of seawater as it gets warmer and widespread melting of land-
based ice (Pittock, 2009, p. 20).

The observed sea level rise and model estimates are in agreement within
the uncertainties, with a lack of significant acceleration of sea level rise
detected during most of the twentieth century. The absence of an observed
acceleration in sea level rise up to the 1990s is due to long time lags in warming of the deep oceans, but there is evidence of acceleration in the last decade,
probably due to rapidly increasing contributions from the melting of land-
based ice in Alaska, Patagonia and Greenland. Studies by US scientists of
twentieth-century drying trends in the Mediterranean and African monsoon
regions suggest that the observed warming trend in the Indian Ocean, which
is related to the enhanced greenhouse effect, is the most important feature
driving these dryings, through its dynamic effects on atmospheric circulation.

Another study shows a tendency for more severe droughts in Australia,
related to higher temperatures and increased surface evaporation. Both studies see tentative attribution of drying trends to the enhanced greenhouse effect
and are pointers to future regional climate changes (Pittock, 2009, p. 14).

Indicators of Climate Change

Nature has its own way of showing us telltale signs of Climate Change –
species getting extinct, shrinking of river catchments, diminishing of the carrying capacity of rivers, and disappearance of frogs, butterflies and honeybees. "As global warming raises the planet's temperature, many species will
lose territory in which to survive" ("Climate Change threatens," 2017, p. 4).

Researchers at the Environmental Management and Policy Research
Institute say that butterflies are the best bioindicators of Climate Change
(Rohith, 2017). Scientists have synthesized currently available evidence and
documented the profound impacts of glacier mass loss on freshwater and
near-shore marine systems. Glaciers cover close to 10% of the Earth's land
surface but are shrinking rapidly across most parts of the world.

Human activities affect climate in many ways. Deforestation and subsequent land use for agriculture or pasture, especially in tropical regions, contribute more to Climate Change than previously thought. Even if all fossil fuel emissions are eliminated, if current tropical deforestation rates hold steady through 2100, there will be a 1.5°C increase in global warming. While CO_2 collected by trees and plants is released during deforestation for converting natural lands to agricultural purpose and other human usage, other greenhouse gases – specifically N_2O and CH_4 – are released. These gases compound the effect of CO_2's ability to trap the Sun's energy within the atmosphere. This contributes to radiative forcing – energy absorbed by Earth versus energy radiated off – and a warmer climate. As a result, while only 20% of the rise in CO_2 caused by human activity originates from land use and land-cover change, the warming proportion from land use increases to 40% once co-emissions like N_2O and CH_4 are factored in ("Deforestation contributes more," 2017).

How can we deal with Climate Change?

The goal of the Paris Agreement on Climate Change, as agreed at the Conference of the Parties in 2015, is to keep global temperature rise in this century to well below 2°C above preindustrial levels. It also calls for efforts to limit the temperature increase even further to 1.5°C. The UN Environment Emissions Gap Report 2017 presents an assessment of current national mitigation efforts and the ambitions countries have presented in their Nationally Determined Contributions, which form the foundation of the Paris Agreement. The report was launched on 31 October 2017, days before countries gathered in Bonn for the annual Conference of Parties of the UN Climate Change Convention.

The report has been prepared by an international team of leading scientists, assessing all available information (UNEP, 2017). According to the report, CO_2 emissions have remained stable since 2014, driven in part by renewable energy, notably in China and India. This has raised hopes that emissions have peaked, as they must by 2020 to remain on a successful climate trajectory. However, the report warns that other greenhouse gases, such as CH_4, are still rising, and a global economic growth spurt could easily put CO_2 emissions back on an upward trajectory.

The implications that were made publicly available in 2015 projected that by 2030, carbon emissions are likely to reach 11 to 13.5 gigatonnes of CO_2 equivalent ($GtCO_2e$) above the level needed to stay on the least-cost path to meeting the 2°C target. One gigatonne is roughly equivalent to one year of transport emissions in the European Union (including aviation). The emissions gap in the case of the 1.5°C target is 16–19 $GtCO_2e$, higher than previous estimates, as new studies have demonstrated.

"The Paris Agreement boosted climate action, but momentum is clearly faltering," said Dr. Edgar E. Gutiérrez-Espeleta, minister of environment and energy of Costa Rica, and president of the 2017 UN Environment Assembly. He further noted, "We face a stark choice: up our ambition or suffer the consequences."

Very often, reporting on Climate Change is apocalyptic, where "doomsday narratives" about what Climate Change will "wreck" serve only to bring about a sense of hopelessness among readers. According to an article published in *The Guardian*, "Feeling hopeless about a situation is cognitively associated with inaction and predicts decreased goal-directed behavior," leading the reader to look the other way. "Climate change adaptation only works when we are hopeful for the future and believe that environmentally vulnerable communities have the agency to act."

As President Tong says:

> [T]oday our country (Kiribati Island) is on the frontline. If and when we fall there will be another frontline, and that frontline will keep moving back until those countries and those people who believe they will not be affected will be affected and that is the reality that we are facing . . . there is nothing more globalized than Climate Change.

Western Ghats – a macrocosm of India's biodiversity

To ensure this study was getting wholesome inputs from all stakeholders, an expert panel was consulted informally throughout the study period.

The informal panel comprised scholars, practitioners, economists and journalists:

1 Dr. Usha Rani Narayana
Professor (Retired), Department of Communication and Journalism
University of Mysore, Mysuru
2 Dr. A. S. Balasubramanya
Former Professor and Chairman
Department of Mass Communication and Journalism
Karnatak University, Dharwad
3 Dr. Lingaraj Jayaprakash
Fellow, McGill University, Canada
4 Dr. Sashi Sivramakrishnan
Professor of Economics, NMIMS-Bengaluru
Documentary filmmaker of "Faces of Kudremukh"
5 Dr. R. Siddappa Shetty
Fellow (Associate Professor)
Centre for Environment and Development

Coordinator, Community Conservation Centres (BR Hills and MM Hills)
Ashoka Trust for Research in Ecology and the Environment (ATREE),
Bengaluru

6   Rohith B. R. (deceased February 2020)
    Assistant Editor, *The Times of India* (*TOI*), Bengaluru

Framing the research question

To analyze Press narratives of Climate Change, print as a medium was chosen for this study. The events that occurred from June 2012 to June 2017
related to the declaration of the Western Ghats, as a UNESCO World Heritage Site, which may have led to an increase in newspaper reporting, and
have thus been listed chronologically later.

The aim here is to understand media's framing of Climate Change in this
time of rapid environmental and social changes and to use the declaration
of the Western Ghats as an ESA by UNESCO as a case to illustrate the representations of media reporting and its relationship to policy and marginalized stakeholders. Hence, this book will contribute to the broader literature
on climate communication and media's role in climate communication with
a critical-geopolitical analysis of the issue.

The analysis of media reports has been broken down in two sections:
the first one deals with the coverage of Climate Change; the second one
focuses on the media representation of the Western Ghats after the subsequent declaration of this ecologically sensitive zone as a World Heritage Site
by UNESCO in July 2012 in the southern Indian state of Karnataka.

Given the urgency of the situation and the need for intersection of three
dimensions of Climate Change – the science, its communication (media)
and policy efforts – this study believes that this intersection will lead to an
informed public who can collectively move toward Climate Action. The
study begins with analyzing media content.

The qualitative and quantitative analyses of 552 articles from two major
Indian broadsheets (*Deccan Herald* and *TOI*) have enabled this study
to assess the presence of typical journalistic frames such as conflict and
dramatization.

Key research questions

1   What sources are evident in statements and media coverage of Climate
    Change in the Western Ghats?
2   What were the primary reasons for the peak in media reporting in the
    study period?
3   What is the relevance of the media coverage to the policy imbroglio cur-
    rently plaguing the Western Ghats?

Frequency of Climate Change reporting

Thanks to recent advances in technology, globally comparable data is now available on the ecological state of the planet and human pressure, forest cover and media reports, especially from newspapers all over the world, on Climate Change (Allan et al., 2017). This allows researchers to analyze trends across the globe.

What is happening on the final frontiers (the Arctic and the Antarctic) does not stay at the poles; it will eventually affect the entire planet. Polar wildlife, ecosystems and indigenous populations are already feeling the impact of Climate Change. Traditional knowledge about the environment, passed down through many generations, is breaking down. Ocean circulation and jet streams cause changes in the lower altitudes, where human population is dense. Hence, media reports on polar bears and ice caps melting at alarming frequencies must be perceived by the media in a more responsible manner (through prominent placement and in-depth coverage), as well as the readers, rather than assuming that the news is irrelevant in relation to the geographical location.

For the period 2000–2017, an increase in the frequency of media reports related to Climate Change can be found. This is linked to the environment summits in Copenhagen (2009) and Paris (2015). World leaders from over 150 countries attended these historic summits and signed treaties to mitigate and fight this global concern. This in turn sparked widespread media coverage.

Analysis of media reports across the world from 2000 to 2017 shows that events like the Paris Agreement and Copenhagen summit create a sudden interest in the media, leading to the appearance of news articles (Anderson, 2009). Following the same rationale, this study uses the argument to emphasize that media reporting surrounding Climate Change is influenced by policy decisions that are an outcome of major environmental summits (such as Copenhagen and Paris).

Using the declaration of the Western Ghats as a World Heritage Site in 2012, by UNESCO, as evidence, the coverage and media representations following this decision in the state of Karnataka have been analyzed. The decision has implications for policy and forest communities, and the state is yet to come to a consensus after constituting two expert committees (Kasturirangan and Gadgil) to guide them in this matter. By earmarking certain zones in the Western Ghats as ESAs, there arises a conflict between development and conservation. In this conflict situation, media reporting plays its own role. The argument that media narratives or media coverage surrounding Climate Change peaks during conflicts and international summits can be directly related to the presence of world leaders, the important announcements they make, that have implications on business and the economy.

However, between 2003 and 2006, there was a steady increase in attention on Climate Change among the opinion-leading newspapers in the

United States, with an even more marked rise in the United Kingdom (Boykoff, 2007; Boykoff and Roberts, 2007). The release of Al Gore's film *An Inconvenient Truth* clearly played a part in generating the rise in media coverage in 2006. Over this period, there was also an increase in coverage in Australia/New Zealand, the Middle East, Asia, Eastern Europe and South Africa, though again this was less steep than in the UK (Boykoff and Roberts, 2007, p. 39). Ironically, there has been comparatively little media coverage of Climate Change in developing countries; yet they are likely to suffer the worst effects (Painter, 2007). In 2006, there was four times as much coverage of Climate Change issues in the UK Prestige Press than there was in 2003. However, in the United States over the same period, coverage increased in the elite national Press by roughly two and a half times (Anderson, 2009; Boykoff and Boykoff, 2007).

In the world coverage of Climate Change, there was a peak noticed in reporting in December 2009. The coverage of Indian newspapers updated through October 2017 (Nacu-Schmidt et al., 2017) tracked newspaper coverage of Climate Change or global warming in four Indian newspapers (*The Indian Express, The Hindu, Hindustan Times*, and *TOI*).

There was a peak in newspaper coverage of Climate Change in December 2009, with *The Hindu* (268), *TOI* (184), *Hindustan Times* (133) and *The Indian Express* (106) leading the list.

The fifteenth session of the Conference of the Parties to the UNFCCC and the fifth session of the Conference of the Parties serving as the Meeting of the Parties to the Kyoto Protocol took place in Copenhagen (UNFCCC, 2017).

The second peak in newspaper coverage of Climate Change was witnessed in December 2016, with *TOI* (219) topping the list, followed closely by *Hindustan Times* (210), while *The Hindu* (175) and *The Indian Express* (118) trailed behind.[1]

On 5 October 2016, the threshold for entry into force of the Paris Agreement was achieved. The Paris Agreement entered into force on 04 November 2016. The first session of the Conference of the Parties serving as the Meeting of the Parties to the Paris Agreement (CMA 1) took place in Marrakech, Morocco, from 15 November 2016 to 18 November 2016 (UNFCCC, 2017).

From a combined coverage of 20 in 2000 to 397 articles in total in 2010, the peak was witnessed in 2009 and 2015, when the combined tally was 700. As of October 2017, *TOI* had the highest coverage with 104 articles, out of the 296 articles by the four dailies.

Dimensions of media reporting

The World Heritage Convention was adopted in 1972 to ensure the world's most valuable natural and cultural resources could be conserved in perpetuity (UNESCO, 1972). The Convention aims to protect places with Outstanding Universal Value that transcend national boundaries and are worth conserving for humanity as a whole. These places are granted World Heritage status, the highest level of recognition accorded globally. A unique aspect of the convention is that host nations are held accountable for the preservation of their World Heritage Sites by the international community and must report on their progress to the United Nations Educational, Scientific and Cultural Organization (UNESCO). Over 190 countries are signatories to the convention, committing to conserving the 1,031 World Heritage Sites listed. Of these, 229 are Natural World Heritage Sites (NWHS), inscribed for their unique natural beauty and biological importance.

The declaration of the Western Ghats as one of the NWHS has been used as a focal point in this study to assess the dimensions of media reporting. We argue that the policy implications and conflict arising from the overlap of development and conservation between the key stakeholders – government, business and public – will form the core competencies of media reporting on this issue.

Economic implications

Media reports are likely to focus on reporting about a crisis, such as decline in production of crops, especially since the Western Ghats are producing some of the key cash crops – namely, coffee, tea, rubber, cardamom, pepper, arecanut. A study on the importance of bee pollination in coffee production in Kodagu by researchers Virginie Boreux and Lavin Biddanda at the College of Forestry, Ponnampet in Kodagu district, reveals that bee pollination increases the number of coffee berries harvested per cluster. The research paper states:

> Bees have a particular behaviour when they collect nectar and pollen: they visit many flowers from different coffee bushes. Thus, many pollen grains of different origins are deposited on the stigmas, and through competition, only the best of these pollen grains manage to fertilise coffee flowers.

This has, in turn, affected the production of honey and the domestication of honeybees, which have been in existence for around 120 million years. Nearly one million tonne of honey is produced worldwide every year, yet the delicate nature of the honeybees – which collect wild flower nectar in order to produce honey – are facing a threat due to the impact of Climate Change. Along the Western Ghats, the honey collection societies of Kodagu, Sakleshpur and Dakshina Kannada are facing a rather uphill task of sourcing honey locally as well as the challenge of catering to a huge demand. An official from the Coorg Progressive Beekeepers Co-operative Society, Bhagamandala, states:

> In the Talacauvery belt, a local tribe called Jenu Kurubas would venture deep into the forests and climb the tallest tree to bring us honey. Today, most of the people in this tribe have migrated to the

cities. It is during the months of March and June when the coffee flowers blossom that we get a good yield of honey,

"We also have a bank of 200 commercial beekeepers and give a subsidy to these farmers to keep a box for the honeybees," he says, while agreeing that the area of operation has shrunk rapidly due to deforestation and deficit rainfall in the region. Further along the Western Ghats is the South Kanara Beekeepers Society, Puttur. An official says:

> Every year, we get close to 30,000 kg of honey from the entire district, mainly from the honeybee which feed on plants such as rubber. There is an increasing demand for honey but we cannot meet the demand as production is less. This is a lucrative business. We provide wooden beehives to the farmers, and have around 10 large-scale beekeepers,

However, the commercial beekeepers are also largely dependent on the ecosystem. The demand for honey has gone up over the years as the world's population has swelled and lifestyle diseases are rising, and the importance of honey for its medicinal value has been promoted.

"Bees have not changed, but the environment they have thrived in for centuries has changed," says Satish Kumar, an entrepreneur in the honey-processing industry. There are many indicators of Climate Change, but a look at the honeybee production patterns will possibly give a whole new insight into the situation that demands our attention, since it is also linked to our very own existence (Ganapathy, 2017, p. 4).

A recent study by Jayakumar, M., Rajavel, M., and Surendran, U., high-lighted the drop in coffee yield in the Western Ghats due to Climate Change, though production and demand for coffee are on the rise in the last decade (2017). Coffee is the world's most valuable tropical export crop. Recent studies predict severe Climate Change impacts on production. However, quantitative production figures are necessary to provide coffee stakeholders and policy makers with evidence to justify immediate action. Similar to other equatorial regions, warming has occurred over much of Tanzania, particularly since 1970 (IPCC, 2007; Craparoa, Van Astenb, Läderachc, Jassogneb, and Graba, 2015).

Recently, there has been increased attention on the substantial rise in nighttime (minimum) temperatures and the effect this has on tropical crops, particularly in India and Southeast Asia (Bapuji Rao et al., 2014; Nagarajan et al., 2010).

Coffee is India's important export crop, generating US$699.67 million in 2015–2016 (provisionally based on export permits from 1 April 2015 to 29 February 2016), growing at a CAGR of 11.05% during the period. The domestic industry has seen a sudden growth in coffee consumption from around 50,000 million tonnes (MT) in 1998 to 115,000 MT in 2011 (provisional estimates), registering a CAGR of 6.09% (Coffee Board of India, 2017). The area under coffee plantations in India has increased by more than three times, from 120.32 thousand hectares in 1960–1961 to 397.147 thousand hectares in 2015–2016. Most of this area is concentrated in the southern states of Karnataka (54.95%), Kerala (21.33%) and Tamil Nadu (8.18%) in the Western Ghats.

The seventh major exporter of coffee in the world, India's commercial plantations of coffee started during the eighteenth century and is the only country in the world where all types of coffee beans are grown under a "well-defined two-tier shade canopy of evergreen leguminous trees" in one of the 25 biodiversity hotspots in the world.

The final crop estimate based on crop harvest data for the year 2016–2017 is placed at 312,000 MT, showing an overall decline of 36,000 MT (–10.34%) over the previous year's, 2015–2016, estimated record crop of 348,000 MT. The reason for reduction in production estimates of 2016–2017 is attributed to the delayed blossom and backing showers coupled with high temperatures (Coffee Board of India, 2017).

Among the states, the final estimate for Karnataka is placed at 221,745 MT comprising 70,510 MT of Arabica and 151,235 MT of Robusta, recording a decline of 4,555 MT (–2.01%) over the post-monsoon estimate of 2016–2017. The production of Arabica has declined by 1,090 MT (–1.52%) and Robusta by 3,465 MT (–2.24%) over the post-monsoon estimate.

Among the districts, the major loss of about 4,290 MT is reported from Kodagu district. This is a cause of concern because the coffee-growing regions in the Western Ghats are labor-intensive, attracting and supporting a large segment of daily wage laborers who come during the coffee-picking season (three months annually) from all over India and even from neighboring Sri Lanka.

This has a huge implication for policy decisions, if research is showing that the decline in yield is due to rising temperatures and decline in rainfall. Another worrying trend is the tourism industry, which occupies small holdings of coffee plantations, where the landowner is unable to cope with the uncertainties and chooses to sell his land to luxury hospitality chains that are willing to pay astounding amounts for an acre of land. In 2003, an acre of land in Kodagu district fetched INR 1 lakh. In 2021, it can sell for anywhere between INR 30–40 lakhs.

Implications for policy

The growing controversy surrounding policy dimension and vulnerability has been discussed in the media extensively. India threatened to pull out of the United Nations' IPCC and set up its own Climate Change body. The Indian government's move was aimed to snub both the IPCC and Dr. Pachauri, following the revelation that his most recent Climate Change report included false claims that most of the Himalayan glaciers would melt away by 2035.

Scientists believe it could take more than 300 years for the glaciers to disappear. Jairam Ramesh, the then environment minister at the center, effectively marginalized the IPCC chairman even further. He announced that the Indian government will establish a separate National Institute of Himalayan Glaciology to monitor the effects of Climate Change on the world's "third ice cap" and an "Indian IPCC" to use "climate science" to assess the impact of global warming throughout the country. He said:

> There is a fine line between climate science and climate evangelism. I am for climate science. I think people misused [the] IPCC report . . . [the] IPCC doesn't do the original research which is one of the weaknesses . . . they just take published literature and then they derive assessments, so we had goof-ups on Amazon forest, glaciers, snow peaks. I respect the IPCC but India is a very large country and cannot depend only on [the] IPCC and so we have launched the Indian Network on Comprehensive Climate Change Assessment (INCCA).

India is an agrarian economy, heavily dependent on the Southwest monsoon and the Northeast monsoon for its agricultural activities. If the 1.3 billion population were to merely rely on the vagaries of monsoon and its amplified erratic behavior in the context of Climate Change, the role of the Press in reporting stories related to the economic implications, human development and environmental concerns must take on a bigger chunk of coverage in its everyday coverage.

Assuming 80% of news in Indian newspapers and broadcast media is related to political coverage and the remaining to crime and entertainment, where does that leave coverage to Climate Change matters – a subject of pressing concern indeed.

Note

1 Here is the description provided on the website about the four Indian dailies, which feature among 52 dailies from nine regions across the globe.

The Hindu – Based in Chennai, *The Hindu* is an English-language Indian newspaper that was founded in 1878 as a weekly. The newspaper became a daily in 1889, and it was the first newspaper to start an online edition in India. Kasturi & Sons Ltd. currently owns the newspaper. Circulation reached 15,58,379 copies from July to December in 2012. According to the Indian Readership Survey in 2012, it was the third most widely read English newspaper in India (after *TOI* and *Hindustan Times*), with a readership of 2.2 million people. Archives through Factiva are available from May 1998 and are updated the same day as publication. Full texts of the articles are available in English.

Hindustan Times – Mahatma Gandhi founded *Hindustan Times* in 1924. The English-language Indian newspaper, set up to oppose the British, began as a source to reach a larger audience through an English readership. Today, the newspaper is run by HT Media. The newspaper reaches nearly 3.7 million readers across India. Archives through Factiva are available from October 1997 and are updated the same day as publication. Full texts of the articles are available in English.

The Indian Express – *The Indian Express* is a daily Indian English-language newspaper that is published in Mumbai by the Indian Express Group. The Indian Express Group is under the chairmanship of Viveck Goenka, and *The Indian Express* is its flagship newspaper. Its online version receives 18 million page views a month. P. Varadarajulu Naidu originally founded the newspaper in 1932. In 1991, following the death of the paper's subsequent owner, Ramnath Goenka, the Goenka family divided the newspaper into two separate companies. In 1999, the northern editions, headquartered in Mumbai, retained and renamed *Indian Express* as *The Indian Express*, while the southern editions became *The New Indian Express*. Today, the two newspapers and the two companies are separate entities. Archives through Factiva are available from May 2007 and are updated the same day as publication. Full texts of the articles are available in English.

The Times of India – Founded in 1838, *the Times of India* (TOI) is an Indian English-language daily newspaper. In 2008, the newspaper was dubbed the world's largest-selling English-language daily, with a circulation of over 3.14 million. Today, TOI is the most widely read English newspaper in India with a readership of 7.643 million. The newspaper is published by Bennett, Coleman & Co. Ltd., which is owned by a Jain family. Archives through Factiva are available from May 1986 and are updated the same day as publication. Full texts of the articles are available in English.

3

COMPARATIVE ANALYSIS
OF TWO INDIAN BROADSHEETS

The forest was shrinking
But the trees kept voting for the axe
For the axe was clever and convinced the trees
That because his handle was made of wood, he was one of
them.

~ Turkish Proverb[1]

"We are the last generation to address Climate Change," is a profound statement that is amplified and reverberated in every major environmental summit today. "Global warming" was a term that was emblazoned across the front page in newspapers across the globe. It became a term that was used and overused and at times misused and misinterpreted by vested interests, and before the world took cognizance of the problem, the cause of "global warming"' resulted in the effect of "Climate Change."

Research around Climate Change is multidisciplinary, involving earth science, environmental science, economics, anthropology, sociology, communication as it seeks to simplify the complexities that have resulted from an amalgamation of pure science and social science. Weart (2012) traces the anthropogenic origins of this phenomenon, which began with the increase in research activity from environmental scientists in the 1970s, who raised doubts about the benefits of human activity for the planet.

Scientific reporting and journalistic reporting have a common thread that binds them and interweaves the reasoning and logic to showcase the two pillars of truth and accuracy.

However, scientific reports have time on their side, while journalists have to race against time. The Press Council of India states in its Code of Ethics:

> [A]ccuracy is critical since important personal and policy decisions may be influenced by media reports. Once the owner lays down the policy of the newspaper for general guidance, neither he nor

DOI: 10.4324/9781003015673-3

anybody on his behalf can interfere with the day to day functioning of the editor and the journalistic staff working under him.

While journalists are bound by their organizational boundaries, the The United Nations Educational, Scientific and Cultural Organization (UNESCO), the United Nations Environment Program (UNEP), the World Meteorological Organization (WMO), the International Telecommunication Union (ITU) and the UNFCCC formulated a set of code of ethics for journalists at the Paris Summit in 2009 (see Box 2).

Box 2 Code of Ethics signed under Paris Agreement for journalists

Objectives

- To strengthen regional and international collaboration of all broadcasting organizations and concerned professional organizations to optimize the quality and relevance of programming and reporting on Climate Change;
- To encourage the production and dissemination of relevant audio-visual content at a local level to give voice to marginalized populations affected by Climate Change;
- To collaborate in raising the skills of broadcast media professionals through training, the exchange of knowledge and best practices and by facilitating access to relevant scientific information;
- To promote opportunities for media professionals to build information-sharing networks; And invite the broadcasting unions and other international associations of broadcasters;
- To encourage their members to make quantifiable commitments to increase the availability of content on Climate Change through the exchange of audiovisual material and the broadcast of programme items at local, national and international levels;
- To develop and promote broadcasting industry standards in environmental management and to urge their members to set quantifiable targets for a reduction in their own carbon footprint.

Emphasizing that an increased public understanding of the urgency of Climate Change is essential to mitigate its negative impacts and to avert human suffering;

Underlining that access to relevant information on Climate Change is vital to sustain a living planet and for the survival of human beings;

Acknowledging that there are significant social, economic and environmental benefits in taking action to combat the effects of Climate Change;

Recognizing that the information provided by broadcast media plays a critical role in stimulating policy debate and in mobilizing knowledge to empower societies to make informed decisions on options for mitigation and adaptation;

Agreeing that dedicated collaboration among broadcast media to share and disseminate Climate Change information that incorporates both global and local perspectives would encourage individuals and policy makers to undertake timely action.

Climate Change and media studies in India

The media are instrumental in shaping public understanding of environmental issues in India (Chapman, Kumar, Fraser, and Gaber, 1997). The role of mass media in shaping public understanding of environmental issues has been well documented in recent years (Burgess, 1990).

Billet (2010) mapped out how nationally circulated English-language newspapers serving an elite readership in India in the period between 2002 and 2007 fortified rather than caused a breakdown of barriers stemming from Climate Change. "Billet's work has drawn out how there have been differences emergent in Indian and US press coverage of anthropogenic climate change over time" (Boykoff, 2011, p. 51).

Nirmala and Arul Aram in a study on how online newspapers in two English and two Tamil newspapers framing environmental news found the coverage to be low in comparison to crime and politics. Articles appeared in the environment section than prominent front-page coverage.

Poornananda (2017), analyzing the coverage of newspaper in Karnataka related to displacement, found that environmental stories in newspapers constituted only a small percentage of all news reported during the two-year study period.

Displacement as one of the environmental issues was the second most covered issue after deforestation. Although Kannada newspapers which have higher circulation and readership in the districts, taluks and villages comparatively carry a higher percentage of displacement news than English newspapers there are no substantial differences between the newspapers in the two languages. No adequate evidence was found to support the indications of the earlier studies that Kannada newspapers are pro-development and English newspapers are pro-environment Poornananda.

(p. 40)

Brüggemann and Engesser (2017) analyzed media content in Germany, India, Switzerland, the United States and the United Kingdom and correlated this with a survey of the journalists of the respective articles. The study found that journalism has moved beyond the norm of balance toward a more interpretative pattern of journalism. Based on findings from other studies (Billet, 2010; Painter, 2011), this study reiterates that Indian media does not seem to contest anthropogenic Climate Change against the backdrop of the four IPCC statements.

Studies on the Western Ghats

The Western Ghats were declared a World Heritage Site by UNESCO in 2012. This research project focused on a five-year analysis of the coverage of media reporting on Climate Change in the period between 2012 and 2017. According to UNESCO's description, the Western Ghats are older than the Himalayas and represent geomorphic features of immense importance with unique biophysical and ecological processes.

The Western Ghats are India's rainforests – a thriving ecosystem in the southwestern peninsula of India, rich in biodiversity, a treasure house of endemic species, pristine and magnificent, and spanning an area of 164,280 square kilometers.

Home to around 50 million people, the Western Ghats consist of 4,156 villages across six states, running north to south over a distance of 1,500 kilometers, with peaks ranging from 1,030 to 2,695 meters above sea level and rainfall ranging from 80 to 320 centimeters on an average. They are the source of major river systems, including the Cauvery, Krishna, Godavari, Palar and Pennar basins.

According to documented reports, the Western Ghats cover barely 5% of India's forest area yet is home to 1,800 species and consists of 27% of all species of higher plants recorded in the Indian region. The site's high-montage forest ecosystems influence the Indian monsoon weather pattern. Moderating the tropical climate of the region, the site presents one of the best examples of the monsoon system on the planet. It has an exceptional level of endemism and is recognized as one of the world's eight "hottest hotspots" of biological diversity. The forests of the site include some of the best representatives of non-equatorial tropical evergreen forests anywhere and are home to at least 325 globally threatened flora, fauna, bird, amphibian, reptile and fish species.

In Karnataka 1,575 villages in the districts of Udupi, Shivamogga, Mysuru, Mangalore, Karwar, Hassan, Kodagu, Chamarajanagar, Chikmagalur and Belagavi come under the Western Ghats ecosensitive area (MoEF, 2013). Several studies carried out in the Western Ghats region enumerate the cause for changes in the freshwater ecosystem during the impoundment of Kali River. The ecosystem of Kali has been developing for so many years, and most of the times it had supported the migrating tribes that have settled along its banks. At present, the age-long harmony is disturbed. The Kali River was enjoying its natural flow through the Western Ghats via several steep and narrow gorges. The run-off-river generation of power is possible during the monsoon, so the turbines have to be kept revolving all around the year by impounding with water from the vast reservoirs behind high-rise dams. The wooded valleys are now being converted into sheets of water.

The increasing population has a direct bearing on increased demands for agricultural land, and this has always had an adverse effect on valuable forests. The high-tension electricity lines riddling the forests, settlement of displaced persons and release of lands for irrigation and hydel projects have a heavy impact on forests.

Rohith (2017) reports on a study "Butterflies as bio-indicators of climate change" indicating that all is not well with our environment. Heightened sensitivity to ecological changes makes butterflies effective agents of Climate Change, according to the report.

Studies related to issues surrounding the ESAs as delineated by the Kasturirangan Report (MoEF, 2013) show that Karnataka has the largest share of ESAs (36.37%), followed by Maharashtra (30.51%) and Kerala (17.59%) (Nair and Moolakkattu, 2017).

Research method

The main objective of this research project is to analyze the news reports on Climate Change in the Western Ghats in Karnataka state. Because the issue of Climate Change cannot be confined to a geographical area, news reports relating to Climate Change across the world are included for analysis.

Scholars have noted that Indian media's climate coverage debates Climate Change not in terms of contrarian versus climate science but as a conflict between traditional CO_2 emitters and emerging economies (Billet, 2010; Brüggemann and Engesser, 2017; Painter, 2011). India being an emerging economy, the fine line to balance tits policy has to thread between conservation and development – and this is rather precarious. Newspapers contribute to this perception of growth, employment and regularly showcase global ratings to show how India and Indian cities fare better when it comes to standard of living and quality of life. If we compare the Indian Print media to the print media globally, the Audit Bureau of Circulation (ABC), the most credible source of listed newspapers, indicates that while the circulation figures for print publications are on the decline in most developed markets, Indian newspapers show an unusual upward trend in circulation.

Circulation rose by 14% and 18% in 2013 and 2014 compared to previous years, slowing to a 12% growth in 2015. During the same years, the United States, the United Kingdom, Australia, France, Germany and Japan saw circulations fall by between 4% and 12%. Only Australia saw a positive growth in circulation numbers for one year, 2014, with 12% more circulations over 2013 (Audit Bureau of Circulation, 2017).

Publishers enroll with the ABC as members and voluntarily provide their circulation figures. ABC certifies these figures after conducting a rigorous audit process through over 90 chartered accountants and audit firms empaneled with it. It also carries out surprise visits to the publication's presses and the studies the markets where it is sold.

ABC certifies circulation figures of its members every six months – for the periods January to June and July to December. It has 967 publications as its members, which include 910 daily and weekly newspapers and 57 magazines and annuals across the country.

Apart from the publications, ABC also has media agencies, ad agencies, government organizations, advertisers for print and the Directorate of Advertising and Visual Publicity of the government as its members. These agencies use ABC figures to plan their advertising spends and marketing plans, as the figures are also available at the granular level of towns and cities in all the states where it has member-publications.

ABC is a founding member of the International Federation of Audit Bureau of Circulation established in 1963. In collaboration with Media Users Research Council, the ABC has set up the Readership Studies Council of India to bring out the annual Indian Readership Survey, the primary survey of print media consumer demographics and the product consumption habits in the country.

ABC, a nonprofit organization that has been certifying circulation figures of member publications since 1948, reported that the average number of copies of print media publications in India went up by 2.37 crores between 2006 and 2016. This translates into a compound annual growth rate (CAGR) of 4.87% over the ten-year period.

ABC's analysis of the print media industry's growth over the last decade shows average copies per day rose from 3.91 crores in 2006 to 6.28 crores in 2016, with the north zone showing the biggest CAGR spike of 7.83%. ABC said in a statement that the print medium in India "is thriving, growing and expanding" in spite of "stiff competition" from television, radio and digital industries.

The growth in print has been powered by Indian languages, the ABC figures show. Hindi grew the fastest (CAGR 8.76%) during the 2006–2016 decade, a finding that ties in with the fastest circulation growth in the north zone, followed by Telugu (8.28%), Kannada (6.40%), Tamil (5.51%) and Malayalam (4.11%). English publications saw a below-average growth over the decade, at a mere 2.87% (Audit Bureau of Circulation, 2017).

Hence, for this study, two prominent newspapers *Deccan Herald* and *TOI* in the study area of Karnataka were chosen as the "media" for the research project. The focus was placed on these two newspapers due to their high daily average circulations as well as influence on smaller newsrooms.

It is important to note that both *Deccan Herald* and *TOI* have many vernacular dailies published from their respective media houses. The voice of the vernacular Press is significant given their inroads into the Indian rural places. Moreover, the English reading population in India is around seven million. However, at the macro level the policy makers and key stakeholders continue to look for the "voice of the masses" through major English-language newspapers. Policy actors routinely monitor these media sources for salient aspects of contemporary public discourse including climate science (Boykoff, 2011).

Using the key words "Climate Change" and "Western Ghats" the e-paper archives database of *TOI* was relied upon to pull comparable samples from *Deccan Herald* and *TOI* for this research. *Deccan Herald* does not make its archives freely downloadable unlike *TOI*. Hence, we had to seek special permission from the library division at the Printers Mysore Ltd., which owns and publishes *Deccan Herald*. We had to visit the library on M.G. Road premises in Bengaluru at a specified time. This constraint was due to the fact that the library was undergoing renovation and also that there were only limited number of machines available for use. Since all the archives are available on Comyan, which is a media management software, and only on the servers of the Printers Mysore Ltd., this was a huge challenge. We were allotted two hours every Saturday to access one machine in the library, and we were informed that we have to pay INR 50 for each article accessed. This highlights the challenges that the researchers face, when data is not made available – in a free and fair manner for the purpose of academic research.

The key words were "Climate Change" and "Western Ghats." Given that the search results (Comyan and *TOI* e-paper) included many articles where Climate Change was merely referred to in passing, the initial database was narrowed down to those articles where it was the core theme. This required a quick reading of all texts to select those where the key words were mentioned not only in the headline but also in a significant portion of the article. Climate Change or Western Ghats and the implications for Climate Change had to be presented as the central theme. A total of 552 articles met at least one of these conditions. This means a total of 522 articles from these two Indian national newspapers (Karnataka edition only) were considered for evaluation.

The *TOI*, published in English since 1838, a broadsheet national daily having a circulation of 3,184,727 (ABC, July–Dec 2016), is owned and published by Bennett, Coleman & Co. Ltd. It is the oldest English-language newspaper in India still in circulation, dating back to its first published edition of 1838.

Deccan Herald, published in English since 1947, is a broadsheet regional daily with a circulation of 269,239 (ABC, July–Dec 2016).

ABC was established in the year 1948. It is a not-for-profit organization and is in continuous operation since the last 69 years. ABC is also the founder member of the International Federation of Audit Bureaux of Certification (IFABC) since 1963. ABC has been continuously certifying circulation figures of member publications every six months, that is, for the audit periods January–June and July–December since its inception. The trend of certified circulation figures by ABC shows that the print medium (member publications of ABC) is thriving, growing and expanding in India in spite of stiff competition from all other mediums, namely television, radio and digital. Publishers enroll themselves as members of ABC to get their circulation figures audited. ABC certifies circulation figures after a stringent audit process through more than 90 empaneled chartered accountants, audit firms. ABC also has a provision for surprise Press and market visits by empaneled firms of chartered accountants; this further strengthens the audit process.

As on date, ABC certifies 910 daily and weekly newspapers and 57 magazines and annuals in India. It is believed that the growth of newspapers in India is directly related to urbanization leading to higher aspirations, heightened interest in buying assets and so forth. Print is growing at an incredible 4.87% increase in CAGR over a ten-year period. About 2.37 crore copies were added in the last ten years accompanied by an increase of 251 publishing centers (Audit Bureau of Circulation, 2017).

Content analysis

The history of content analysis has gone through different phases. Various approaches have been used for text analysis, including graphological procedures and dream analysis by Sigmund Freud. This approach found its way into linguistics, history, sociology and psychology. After much refining and use of contingency analysis and computer-coded programs, qualitative content analysis has also been criticized for being superficial without understanding the context and latent content.

There are different levels of content in a given text for analysis. At the surface is the basic content that gives the main idea and sets the theme of the text. If you look deeper and probe into the context of the text, you will be able to arrive at the latent content (Becker and Lissman, 1973).

Qualitative content analysis is an alternative to the more common quantitative content analysis; the latter focuses on frequencies of themes or terms that are found in a sample of texts. The goal of quantitative content analysis is to uncover manifest meanings within a text. Such significance helps the researcher to identify any rules or patterns in the usage of terms or themes (Krippendorff, 2012). In contrast, the rationale behind qualitative content analysis is to uncover any underlying meanings found in a text that cannot be figured through the counting of themes or terms (Krippendorff, 2012). Merely counting a theme or term, as in quantitative content analysis, is not enough to understand any latent patterns developed within texts. Essentially, researchers utilize qualitative content analysis to uncover any underlying meanings that are conveyed through the text.

Qualitative content analysis is typically accomplished through one of two processes: a close reading of relevant materials by the researcher and the use of coders. In the first case, the researcher essentially reads each text and notes important elements that fit preexisting categories or elements that might give rise to emergent categories (Altheide and Schneider, 2013; Krippendorff, 2012).

The act of engaging in the reading alone can make for relatively quick and efficient qualitative content analysis. However, this process of qualitative content analysis has two serious flaws: researchers may miss important details or their agenda or research questions may skew their observations. To remedy these flaws, researchers may enlist the aid of additional "coders" in their analysis of texts. The use of coders can be advantageous, as the researcher can demonstrate that multiple readers arrived at similar conclusions. In order to effectively use coders, however, the researcher must be able to establish intercoder reliability, which demonstrates that the researcher and coders in fact came to the same conclusions (Neuendorf, 2002). To establish intercoder reliability, the researcher first needs to thoroughly train the coders about the reading process: What categories will they be searching for in the content as they engage in the reading? What elements should they be searching for within the texts? Such training is essential; otherwise, the researcher and coders may very well look for vastly different things within the text, and the results would be skewed. Once the coders have been trained, they may conduct their reading.

Phillip Mayring (2000) explains that qualitative content analysis is an important tool available to researchers so that they might untangle the concrete and underlying meanings that govern rules, patterns and relationships within texts. He claims that there are two modes for the application of qualitative content analysis: deductive category application and inductive category development. The deductive category analysis is similar to Margrit Schreier's (2012) mode of qualitative content analysis, which has roots in discourse analysis. In this deductive form of qualitative content analysis, the researcher uses categorical frames that have been prepared prior to the analysis to examine a text; the goal is to search for elements of a text that can fit within the preexisting categories or frames. These categories are not concrete terms, as described earlier in reference to quantitative content analysis. Instead, the researcher develops categories that can emerge in a variety of different ways.

For instance, a researcher could search for "activism" within a particular text. Activism is not clearly defined by any one term or a set of terms and can be understood in different forms (e.g., street protest, culture jamming, hacking, wearing T-shirts with political messages); there are different ways of conceptualizing activism. The researchers could utilize a definition of activism as a frame to use to seek out examples of activism within a text; a frame constructed from the methodological position of dialectics or phenomenology would provide some parameters but would also be quite broad. The researcher, or trained coders, can read throughout various texts, looking for forms of activism that fit within the particular frame established at the onset of the project. As the text is examined, readers can identify and note instances of activism within the text. The readers do not merely count the number of times that the category emerges within the text but examine how the category is utilized. In the "activism" example, one or more readers could note forms of activism evoked within the text, as well as the tone and language used around those forms. The readers could also note the uses of characters or settings used in the descriptions of activism. Ultimately, the frame provides multiple examples of activism that exist within the text; the reading and use of the frame allow for latent meanings to emerge. The researcher can examine those examples to find meanings associated with activism through forms, descriptions, tone and other uses of language within the text.

In this study, we have used the inductive category development (Mayring, 2000) and created categories based on our research questions. Each newspaper was examined across a five-year period. Individual newspaper items were selected for analysis using electronic database (described earlier) and the keywords "Climate Change" or "Western Ghats" were used.

Each item was then read to exclude those that made only passing reference to Climate Change issues. The remaining articles ($N = 552$) were subjected to content analysis using a coding schema designed to capture use of voice (positive, negative quotation and paraphrasing of sources from various backgrounds), claims (regarding specific risks and benefits, impacts, motives), narrative devices (metaphors, parallels) and frames (such as green capitalism, political conflict, fatalism). The codes were determined partly with reference to similar studies conducted in Europe and the United States (e.g., Antilla, 2005; Boykoff and Boykoff, 2004; Carvalho, 2007) and also inductively from a preliminary reading of approximately 250 articles in the sample. The strength of structured content coding is that it standardizes the articles and thus allows for statistical analysis of tendencies, variance and associations (Young and Dugas, 2012).

The coding was performed by two coders. Intercoder reliability was tested using a Cohen's kappa test on a subset of 100 articles, following two iterations of a simple percent agreement methodology; intercoder reliability achieved 89% (Lombard, Snyder-Duch, and Brack, 2002, p. 593; Neuendorf, 2002), which is considered high for this test (see Table 3.0 for an overview of the sample).

To measure the content, the following unit of analysis formulae has been applied:

Unit of analysis = one news article

Table 3.0 Overview of sample for the five-year study period (2012–2017)

Number of articles in Deccan Herald	Number of articles in TOI	Total (N)
310	242	552

Table 3.1 Overview of model

1 Collect relevant news articles
News accounts about a policy problem are collected using online databases. Studies show news media reports both reflect and influence public opinion and policy making.
2 Code news articles
Through our categorical content system we:
  • Identify specific arguments relevant to a policy problem and solution to solve it
  • Code those arguments for direction relative to goals and/or principles of a specific policy
  • Describe the substance of those arguments
  • Identify stakeholders linked to those arguments
  • Match arguments to one or more dimensions of the policy process
3 Using coding results to inform elements of the policy sciences analytic model
Analyzing empirical coding data and content of coded arguments enables description of:
  • Social process – (particularly) participants, perspectives, situations, base values and strategies
  • Decision process – (particularly) intelligence gathering and promotion
  • Trends – in each of our content analysis categories

Source: Howland, Becker, and Prelli (2006).

The study period was heavily built-upon the policy-relevant news media accounts. Hence, a categorical system of content analysis suited the study method in the task of analyzing and describing policy-relevant news articles. Howland et al. (2006) proposed a model to analyze and describe policy-relevant media articles (see Table 3.1).

Coding with the following categorization:

1 What sources are evident in statements and media coverage of Climate Change in the Western Ghats?
2 What were the primary reasons for the peak in media reporting in the study period?
3 What is the relevance of the media coverage to the policy imbroglio currently plaguing the Western Ghats?

Pearson's chi-square test was performed to examine the significant association between *TOI* and *Deccan Herald* related to the variables of the study. The relation between these variables was significant in most cases.

Framing, news and public policy

Often considered a fractured paradigm (Entman, 1993), framing is unique in and empowered through its diverse ability to consider cultural, social, political and historical dimensions within and among news storytelling (D'Angelo, 2002; Van Gorp, 2005), revealing what is truly being communicated through news articles. Given India's richly diverse multiethnic communities, this kind of rich theoretical analysis is needed. Framing has the potential to significantly and critically influence citizens' evaluation of issues (Han, Chock, and Shoemaker, 2009; Han and Wang, 2012), affecting how they think about and act upon the issues that confront them (Kuypers, 2002). Further, analyzing framing across news stories helps identify journalism's role along with political elites and governments in fostering official political discourse regarding (inter)national organizations' workings and operations as they unfold and evolve over time (Entman and Rojecki, 1993; Snow and Benford, 1988; Rosas-Moreno and Ganapathy, 2021).

Given this theoretical framework, the main research question for this study is:

Main RQ: *How has the Indian national press framed environment-related news?*

To investigate the framing of politics as a strategic game in Climate Change, two variables were used. The coders were asked to indicate (yes/no) whether the news article "deals extensively with the rights of forest communities, governing negotiations between forest officials, or raising awareness among forest communities about the risks of Climate Change," and whether it "deals extensively with politicians' or parties' taking a stand with elections in mind." To investigate the framing of Climate Change and policy-relevant decisions, three frames were coded. The frames asked: Does the story raise relevant issues and "knowledge about policy in existence," "contribute to the implementation of policy" and "contribute to raise awareness among forest communities and their attitude toward the policy"?

Journalists and media groups have a critical role to play in explaining the cause and effects of Climate Change. To assess this role that the two newspapers displayed over a five-year period, the two frames related to the role of media: Does the news story cast the media as capable of addressing important problems faced by the region and country, through the communication of its Climate Change–related reporting? The frames were analyzed through the following questions: "Does the news report feature crusaders and the common man who is working towards sustainability?" "Does the news report present evidence and arguments to convince the reader to spur them into act and be a part of the solution?" "Does the news report focus on the context of major environment summits?" "Does the news report create a sense of fear about Climate Change and its consequences?"

This book also seeks to frame the challenges that journalists face when reporting on this topic. This frame was analyzed by asking, "Does the news report seek to bring about immediate behavioral change and persuade the reader to do this?" "Is the news report restricted by cultural and social norms in India?" and "Is the news report restricted to reporting on the issue if the source is a celebrity, community leader or government spokesperson?"

An initial typology helped organize and keep track of the data for basic content analysis assessment (see Appendix A).

In terms of the quantity, the data indicated that news coverage had substantially increased in the years (2013 and 2014) following the listing of the Western Ghats as a Natural World Heritage Site (NWHS). Although there were only a total of 259 Climate Change stories (see Table 3.2) in 2013 and 2014, this makes for 47% of the total articles ($N = 552$) in this study sample.

However, the number of articles from June to December 2012 (113) accounted for a considerably large proportion of coverage, indicating that the media seemed to pay significant amount of attention after the declaration was made by UNESCO. This suggests that Climate Change and its related impacts had become an increasingly prominent issue in Indian media, particularly in Karnataka.

Table 3.2 Descriptive statistics ($N = 552$)

Variables	Frequency (%)
Print Media	
TOI	242 (44)
Deccan Herald	310 (56)
Total (*N*)	552
Year	
2012	113 (20.5)
2013	133 (24)
2014	126 (23)
2015	83 (15)
2016	73 (13)
2017 (January–June)	24 (4)
Article types	
News reports	326 (59)
Op-eds (editorial, columns, analytical essays)	29 (5)
Approach	
Climate Change in a positive tone	241 (44)
Climate Change in a negative tone	167 (30)
Focus	
Climate Change as threat to the Western Ghats	271 (49)
Climate Change as threat to forest community	24 (4)
Climate Change in context of policy relevance	251 (45)

Note: Values inside the parenthesis represent percentage of N.

News sources

By examining the sources, readers who are knowledgeable can accurately tell the slant of the news article and the details of the news even without reading the entire article. Journalists are often led by news sources to a particular story, helping the media to set the agenda. Sources are also capable of providing interpretative frameworks, often enumerating to the reporters what the key talking points will be in a story (Entman, 1993). In this manner, news sources can exert significant influences on news content by functioning as an agenda setter or news framer.

Coders first determined the topic of each news article. News sources were then coded by looking at whether each of the following was either referenced or quoted:

1   Law enforcement or official document
2   Legal agency or representative/document
3   Nongovernmental agency or NGO representative
4   UN
5   Citizen
6   Environmental expert

The following research question was put forth:

RQ1: *What sources are evident in statements and media coverage of climate change and articles related to the Western Ghats?*

Table 3.3 shows the classification of news sources with respect to the two newspapers, *TOI* and *Deccan Herald*. The results show that there were clear and significant ($X^2 = 55.653$; df = 5; $p < .000$) trends. A total of 239 (43%) articles in the sample ($N = 552$) quoted or referenced the main source as government officials or law enforcement agencies. These findings may support the idea that Indian media have relied heavily on established routine sources – government officials and lawmakers in this case – in their news coverage of Climate Change.

Table 3.3 indicates another major and credible source for the media in reporting serious global concerns such as Climate Change, citing or referencing environmental experts and their research, appearing in a total of 141 out of the 552 articles analyzed (25%).

Journalists reporting about scientific terminology and ecological issues have always turned to experts to guide their reports – to "balance the bias" and provide sound evidence. "Institutions like CES in Indian Institute of Science are good sources for us as they have considerable expertise with years of research carried out in the Western Ghats."[2]

The results (Figure 3.0) indicate that after official sources and environmental experts, the third major news source for journalists are the public. *Deccan Herald* used citizens as the third major news source in 58% of the

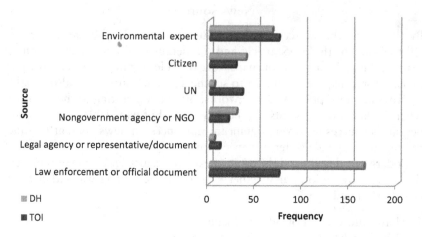

Figure 3.0 Main source for the Climate Change articles in *Deccan Herald* and *TOI*

Table 3.3 Classification of news sources in *TOI* and *Deccan Herald*

News source		Newspaper		Total
		TOI	Deccan Herald	
Law enforcement or official	Count	74	165	239
document	% within Source	31.0%	69.0%	100.0%
Legal agency or	Count	11	5	16
representative/document	% within Source	68.8%	31.2%	100.0%
Nongovernmental agency	Count	20	29	49
or NGO	% within Source	40.8%	59.2%	100.0%
UN	Count	35	5	40
	% within Source	87.5%	12.5%	100.0%
Citizen	Count	28	39	67
	% within Source	41.8%	58.2%	100.0%
Environmental expert	Count	74	67	141
	% within Source	52.5%	47.5%	100.0%
Total	Count	242	310	552
	% within Source	43.8%	56.2%	100.0%

total of 67 articles analyzed, while *TOI* used citizens as the third major news source in 42% of the total of 67 articles analyzed.

News sources from citizens featured eco-warriors, environmentalists and citizen initiatives in areas such as sustainability, renewable resources, and so forth. Prominent headlines where journalists used citizens as the main news source include– "Biking across the blue hills," "Man in Khakhi with green heart," "Chasing a wild dream to strike eco balance," "Engineers go wild, pursue passion," "Students plan India's strategy for Paris meet."

Table 3.3 indicates another interesting trend when comparing the coverage related to news sources in these two dailies. *TOI* used the UN as a news source in 87.5% of the total of 40 articles in this categorization. This can be explained in the context of the extensive reporting on the COP 21 held in Paris from 30 November 2015 to 12 December 2015, during which media coverage also peaked.

Notes

1 Quoted from the page "West Asian Fable", on the Democratic Underground website, www.democraticunderground.com/125616593
2 Personal interview with Rohith B. R., environment journalist, *TOI* Bangalore.

4

CLIMATE – MEDIATIZATION OF PRESS NARRATIVES

A parallel realm that can be compared with media representations of Climate Change is the reporting of pandemics. Scholars have argued that as news of an outbreak such as Ebola in 1995 and H1N1 avian flu are reported, the media representations of the emerging narrative begin with alarm and quickly move to reassurance while framing the narrative to contain the public in quarantine zone and assuring them of the public health system. A sample of stories from five US newspapers, broadcasts that appeared in national network news sample and a collection of internet and social media discussions led scholars to explore how, in the weeks following H1N1's emergence, public health officials attempted to both sound the alarm and manage the reaction. They termed this as the bio-mediatization and bio-communicability process of shaping and coproducing this narrative where government, experts and journalists shaped the discourse (Briggs and Hallin, 2007, 2016; Hallin, Brandt, and Briggs, 2013).

While examining the key Research Question of our study, we looked at – What were the primary reasons for the peak in media reporting in the study period?

"The volume of press coverage provides an important indication of the attention given to an issue over time" (Carvalho and Burgess, 2005, p. 1461). During the period of data collection, we plotted a graph to show the distribution of 552 articles over the five-year study period. Surprisingly, the peak in reporting occurred during two main phases when media attention to Climate Change and the Western Ghats rose significantly. To interpret the patterns of coverage produced by the narrative frequency analysis, a look more closely at the content of coverage during two periods in which there was the greatest attention to Climate Change revealed two peak periods. The first of these periods was in 2013, when the aftermath of the contentious policy issues pertaining to the Western Ghats became a matter of public concern. The second critical period was seen in November 2014, which led up to the Paris Summit in November 2015 (see Figures 4.0–4.2).

Researchers have found similar peaks in other studies conducted for the UK, the US and Japanese media. Media coverage peaked in February 2007,

DOI: 10.4324/9781003015673-4

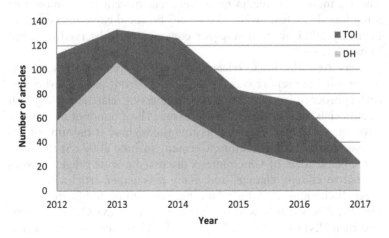

Figure 4.0 Distribution of newspaper articles from 2012 to 2017 in *Deccan Herald* and *TOI*

the month that *An Inconvenient Truth* won two Oscars, Al Gore was nominated for the Nobel Peace Prize and the IPCC released its Fourth Assessment Report with the strongest wording yet about Climate Change. February 2007 also more or less marked the high point of public attitudes on Climate Change. The second critical period is November to December 2009, when the "Climategate" scandal concerning leaked email exchanges among climate scientists and the Copenhagen Climate Summit of world leaders created the decade's most intense spike in news coverage (Anderson, 2009; Carvalho and Burgess, 2005).

A review of the 10-year trend in newspaper coverage of global warming issues revealed that mass media coverage of global warming increased slightly overall before January 2007 and then increased dramatically from January 2007. The increasing Japanese media coverage of global warming issues was driven largely by international events involving the United States, not by all events outside Japan (Sampei and Aoyagi-Usui, 2009).

> This trend in Japanese mass-media coverage is different from that in the UK. Boykoff (2007) investigated the numbers of articles on climate change in three prestigious newspapers in the UK by month from 2003 to 2006. He found a steady increase in coverage in both countries, in contrast to the up-and-down cycle of mass media coverage in Japan around the same period. We rationalized these differences among nations as follows. The Third Conference of Parties to the UN Framework Convention on Climate Change (UNFCCC) in 1997, at which the Kyoto Protocol was adopted, was held in Japan.

The Japanese mass media gave this event quite a large amount of attention. Therefore, when the Kyoto Protocol came into effect in February 2005, more newspaper coverage was observed in Japan than in the UK.

Every time the IPCC released a report from one of the working groups, the report was covered on the front page. During our review process, we found that the numbers of related articles, such as editorials and features, also increased. The amount of newspaper coverage on global warming became the highest at the time of the G8 summit at Heiligendamm, Germany in June 2007. At the summit table, leaders of G8 countries discussed a wide range of issues related to climate change, and many newspaper articles reported these discussions. Al Gore's film was released in mid-September in the UK; this was followed by the Stern Review (31 October 2006) and then UNFCCC COP 12 in Nairobi. These results showed that an increase in media coverage of global warming had an immediate influence on public awareness of global warming issues, but this effect did not last for more than a month.

(Sampei and Aoyagi-Usui, 2009, p. 205)

Previous academic research on the coverage of Climate Change in the Indian media suggests three trends: (a) a growing amount of coverage in the English-language Press since 2007; (b) a dominant framing of the issue as reflective of the world's north–south divide; and (c) a virtual absence of climate skepticism (Billet, 2010; Brüggemann and Engesser, 2017; Jogesh, 2011; Painter, 2011, p. 80).

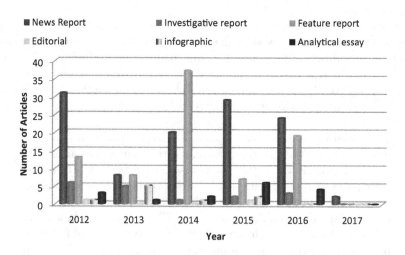

Figure 4.1 Distribution of newspaper articles from 2012 to 2017 in *TOI*

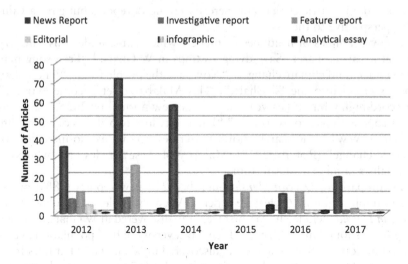

Figure 4.2 Distribution of newspaper articles from 2012 to 2017 in *Deccan Herald*

Table 4.0 Type of articles related to Climate Change from 2012 to 2017 in *Deccan Herald* and *TOI*

Article-type	2012 (n = 113)	2013 (n = 133)	2014 (n = 126)	2015 (n = 83)	2016 (n = 73)	2017 (n = 24)
News report	66 (20)	79 (24)	77 (24)	49 (15)	34 (10)	21 (6)
Investigative	13 (37)	13 (37)	1 (3)	3 (9)	4 (11)	1 (3)
Feature	24 (16)	33 (22)	45 (30)	18 (12)	30 (20)	2 (1)
Travel	4 (100)	0	0	0	0	0
Editorial	2 (67)	0	0	1 (33)	0	0
Infographic	1 (11)	5 (55)	1 (11)	2	0	0
Analytical essay	3 (13)	3 (13)	2 (9)	10 (43)	5 (22)	0

Note: Values inside the parenthesis represent percentage of *n*.

The results show that there were clear and significant (X^2 = 40.262; df = 6; p < 0.000) trends (Table 4.0) across the two newspapers analyzed. A majority of articles (59%) consist of news reports, which is a total of 326 of the sample size (N = 552). This evidence is consistent with the previous finding that the majority of the news source in the media reports came from official sources. A comparative study between *TOI* and *Deccan Herald* show that feature articles (having higher word limit and hence occupying more space in the newspaper) are more prominently included in the media coverage of *TOI*. A feature article can be in the form of a conservation-related topic consisting of 450-plus words. Of the total 152 feature articles, *TOI* published 55% articles that not merely were news reports but gave a valuable perspective to the issue.

A few examples of headlines from *TOI* which came under this category of feature articles are: "Save dying swamps in W Ghats," "Frog song may help understand climate change," "Now save the purple frog," "6 new snail species crawl into the W Ghats," "Has Malabar civet become extinct?" "Record high temp of last year may be the new normal by 2025," "Species can't cope with pace of warming," "Earth may lose two-thirds of its wildlife by 2020: WWF," "Research shows butterflies can act as bio-indicators of climate change" (Rohith, 2017). Majority of these articles had bylines by journalists familiar with this topic.

In comparison, *Deccan Herald* published 212 new reports (38%) of the total sample and only 68 feature articles. A few examples of headlines from *Deccan Herald* include "India's first tree crab discovered in W Ghats," "4 new frog species found in W Ghats," "Rich bio-diversity of bird population in coffee plantations," "Burrowing snake discovered in W Ghats," "Bat tales from W Ghats," "W Ghats helped evolution of two birds," "Variety of fish on verge of extinction" and so forth.

Another interesting finding is that *TOI* carried nine infographics. The importance of visuals in disseminating critical information has been well researched as they carry far more impact in conveying the message effectively (see Figure 4.3).

A comparative study about the prevalence of climate-skeptic voices in the print media in six countries: Brazil, China, France, India, the United Kingdom and the United States found that over 40% of the articles had voices included in the opinion pages and editorials as compared to the news pages. But the print media in Brazil, China, India and France had relatively fewer such pieces than those in the United Kingdom and the United States (Painter, 2011).

This finding has been proved in this study, as there were only three editorials and 23 opinion pieces (analytical essays) across both the dailies. The number of investigative reports (35) across the two newspapers is also remarkably low. Given that the peak in media coverage was due to the conflict in policy issues concerning the Western Ghats, the media has failed to realize its role as the watchdog of society. There was immense scope for the media to use this opportunity to publish more opinions and editorials on this issue, instead of merely reporting the official version of the problem.

Figure 4.3 Infographic in *TOI* chronicling the dispute related to policy in the Western Ghats

Scope of media coverage

The main topic of a story, or the central idea upon which it is based, provides meaning to a sequence of information presented in the story. The organization of this idea serves as a function to package an issue in a certain way, suggesting what the issue is all about. A Climate Change story, for example, can be organized as a policy issue, highlighting the lack of adequate regulations that will address the problems related to deforestation or loss of habitat for wildlife. The primary topics can therefore affect reader's perceptions of where the issue is taking them, shaping the public discourse of the approach toward the issue (Entman, 1993).

The results show that there were clear and significant ($X^2 = 74.611$; df = 5; $p < 0.000$) trends (Figure 4.4). The largest share (39%) of articles in both the newspapers was in the range of 301–400 words. The word count of a news article helps in understanding the amount of prominent coverage given to this issue through the editorial policy of the newspaper. A news report consisting of a Press release or a "he said/she said" narrative does not go beyond the 400 words limit. As newspapers struggle to cope with increasing production and staffing costs and costs related to technological upgradation of their machines and quality of newsprint, they are trying to maintain a firm balance between the ratio of advertisements and news articles on their pages. *TOI* is a market-driven daily with advertisements splashed from the power jacket to the back page, while *Deccan Herald* takes a conservative approach maintaining the 30:70 (advertisement: news ratio) formula. *Deccan Herald* (see Figure 4.4) carries more number of articles in the 301–400 and 201–300 word ranges.

"Journalists are key to the setting of agendas and focusing public interest on particular subjects, which operates to limit the range of arguments and

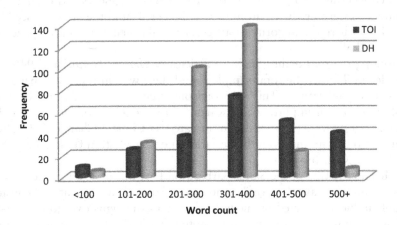

Figure 4.4 Coverage of articles based on word count in *Deccan Herald* and *TOI*

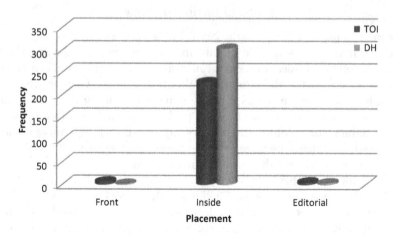

Figure 4.5 Placement of news articles in *TOI* and *Deccan Herald*

perspectives that inform public debate" (Happer and Philo, 2013, p. 321). This agenda-setting function is aided by the prominence given to the placement (see Figure 4.2) of a news article. There is a correlation between the source of an article (official, environmental expert, UN, etc.), type of article (news report, investigative report, editorial, analytical essay, etc.), word count of an article (<100, 101–200, 201–300, 301–400, 401–500, >500) and placement of an article (front, inside, editorial page) that will lead this study to argue about the frames that will eventually emerge in the media's representation of Climate Change.

The Indian media attaches more importance to the placement of news related to crime and politics on the front page and shows less meaning to matters concerning the environment. This is evident in the findings (see Figure 4.5), where the majority of articles are published in the inside pages of the two dailies.

Among the two dailies, 98% of articles in *Deccan Herald* and 94% of articles in *TOI* of the total sample found their way in the inside pages of these two publications. This is a worrisome trend, showing that media does not attach much importance to warrant front coverage or analytical (editorial page) coverage toward issues related to Climate Change and policy implications resulting from the declaration of the Western Ghats as an ESA by UNESCO in July 2012.

Media coverage of the Santa Barbara oil blowout in 1969 in the United States and the "Save the Reef" campaign in Australia documents the considerable influence exerted by media and media campaigns over long periods of time. Foxwell-Norton and Lester noted that in the "Save the Reef" campaign in Australia (2017), "the role of the mainstream media and individual sympathetic journalists was pivotal in bringing scientific knowledge and awareness that elite opinions were contested in the centre of public debate" (p. 574).

Though there was an intersection of political factors with scientific interests in the Western Ghats issue, which was contributing to the critical discourse at that moment, only a small proportion of articles (8) in both the dailies were carried on the front page and there were only two editorials by both the dailies after the government of Karnataka rejected the UNESCO tag of the Western Ghats. This indifference by the media to consistently put pressure on the government is in sharp contrast to similar factors that influenced considerable media coverage leading to pressure on policy makers and business houses in other countries, studied by researchers (Boykoff and Boykoff, 2007; Foxwell-Norton and Lester, 2017).

Earlier studies have demonstrated that media narratives were framed by crisis, disaster and "doomsday" frames. To assess how journalists report on Climate Change and how they deal with denial, journalistic practices of (1) giving disproportionate voice to contrarians and (2) challenging the Climate Change consensus have been used as an analytical framework. Indian media stood out in a study among the US, the UK, German, Chinese and Swiss media for a total lack of challenge of the four IPCC statements (Brügge-mann and Engesser, 2017).

The tone toward reporting on Climate Change and the heritage status accorded to the Western Ghats in this study period show that both the dailies (see Figure 4.6) consistently published articles showing a positive tone toward conservation efforts, renewable energy and policy decisions that would result in mitigation efforts and lead toward Climate Action.

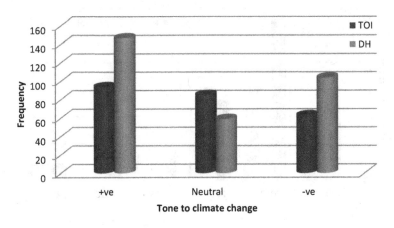

Figure 4.6 Tone of Climate Change reporting in *TOI* and *Deccan Herald*

The results show that there were clear and significant (X^2=18.317; df=2; $p < 0.000$) trends (see Figures 4.7 and 4.8). A distribution of these articles that was analyzed across the study period of five years indicates a strong focus on conservations and the heritage tag that was given to the Western Ghats in 2012. In *TOI*, a large number of articles showing a positive tone were seen in the years 2012 and 2014 (see Figure 4.7). In 2016, however, this editorial stance and media reporting shifted from positive to neutral. This trend points toward a definite shift in the publication's preference from being an advocate to a champion for conservation efforts to showing

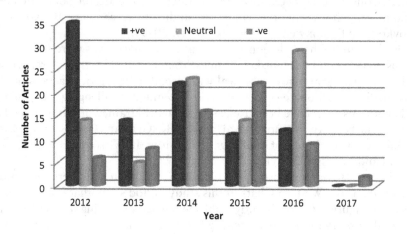

Figure 4.7 Tone of Climate Change reporting in *TOI*

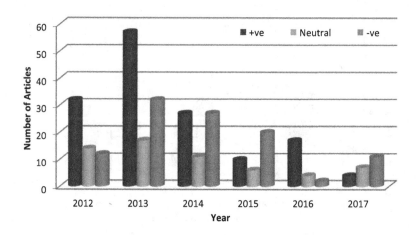

Figure 4.8 Tone of Climate Change reporting in *Deccan Herald*

a "balance" with regard to issues concerning Climate Change and policy implications for the Western Ghats.

Deccan Herald was on the other end of the spectrum, showing a high volume of positive articles in the year 2013 and reversing this trend by showing a negative trend in 2017. This can be attributed to the large number of articles in *Deccan Herald* that are typically news reports, as evidence shown in this study, which do not go beyond the official version. Lately, the Forest Rights Committees and various vested groups have been raising voices against the ESA tag, and *Deccan Herald* had been carrying these reports (in less than 300 words) in 2017 (up to June).

The analysis of these two dailies clearly shows that the Indian media is "sitting on the fence" and merely reporting based on official sources or new scientific research studies that emerge from the scientific community. To come to a conclusion based on the analysis of which of the two dailies are better in media coverage and representation of the issue does not serve the purpose of this study. The main objective was to distinguish the exact nature and mechanisms through which media impacted policy and conservation efforts. In doing so, the study identified the key social actors and media outlets, as well as the interests driving this spike in coverage.

The analysis of media coverage shows that media reporting on Climate Change is still in its infancy, especially in the context of the Western Ghats. The impact of media reporting on conservation and policy is limited in the media, but definitely it exists.

Another key research question we posed – *What is the relevance of the media coverage to the policy imbroglio currently plaguing the Western Ghats?*

To investigate the framing of Climate Change and policy-relevant decisions, three frames were coded. The frames asked:

1　Does the story raise relevant issues and "knowledge about policy in existence," "contributing to implementation of policy," and "contribute to raise awareness among forest communities and their attitude towards the policy?"
2　Does the news story cast the media as "capable of addressing important problems faced by the region and country, through the communication of its climate change related-reporting?"

　　i　"Does the news report feature crusaders and the common man who is working towards sustainability?"
　　ii　"Does the news report present evidence and arguments to convince the reader to spur them into act and be a part of the solution?"
　　iii　"Does the news report focus on the context of major environment summits?"
　　iv　"Does the news report create a sense of fear about Climate Change and its consequences?"

3   The study also seeks to frame the challenges that journalists face, when reporting on this topic. This frame was analyzed by asking, "Does the news report seek to bring about immediate behavioral change and persuade the reader to do this?" "Is the news report restricted by cultural and social norms in India?" and "Is the news report restricted to reporting on the issue is the source is a celebrity, community leader or government spokesperson?"

According to McCombs (2005), the media tend to selectively use certain attributes over others when presenting an issue. In this manner, the selective use of attributes can affect the audience's evaluation of an issue by shaping their understanding of what the reasons to either support or oppose the issue can be.

As many studies have affirmed, content analysis as a methodology has a place in informing policy decisions, and news reports are an ideal source of data rich in context and consequence for both public opinion and policy making (Howland et al., 2006).

The stakeholders identified in this study were (1) law enforcement agencies or official document,[1] (2) legal agency or representative/document, (3) nongovernment agency or NGO representative/document, (4) the UN, (5) citizen and (6) environmental experts.

The key stakeholders that emerged were (1) and (6), pointing toward the relevant issues highlighted in the media through the channel of official sources and with proven research and scientific references from the environmental experts.

"Western Ghats get heritage tag finally" was the headline which was splashed on the front page of *Deccan Herald* on 3 July 2012. This was followed by a few more front-page articles with headlines such as "Heritage tag should boost conservation: Experts" and "Govt opposes UNESCO tag."

The coverage given by *Deccan Herald* toward the Western Ghats heritage tag was 74.2% of the total articles in the sample, showing that the regional newspaper had set its editorial stance very clearly on the issue.

The results show that there were clear and significant (X^2 = 83.458; df = 3; p < 0.000) trends (see Figure 4.9). Following the opposition by the government of Karnataka, *Deccan Herald* ran an editorial on 28 July 2012 with the headline "Save the Ghats":

> The resolution passed by the Karnataka Assembly rejecting UNESCO's heritage status is shocking. . . . It reveals the short-sightedness and small-mindedness of our political class. This is a badge of honour that Karnataka should embrace with pride. . . . the people of Karnataka have welcomed the honour accorded to their mountains. Not so its politicians, who are complaining that the status will not bring any money. The debate on the heritage status in the assembly showed one legislator complaining about the "pittance" that will come with the UNESCO status and were up in arms that "development

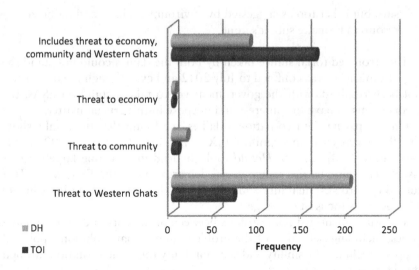

Figure 4.9 Genre breakdown of articles covering the Western Ghats in *Deccan Herald* and *TOI*

projects" will be halted by the heritage status. The theatrics visible in the Assembly would have been (on Wednesday) comical if it were not for the fact that many legislators were being downright disingenuous on several points. Heritage status will not stand in the way of tribals accessing minor forest produce. However, it will stop the rampant levelling of mountains for mining and road construction and felling of trees . . . and it is in stout defence of such activity that legislators voted against the UNESCO tag. . . . the debate and unanimous vote in the Karnataka Assembly reveals that the Western Ghats need protection from the advice of the political class.

TOI ran an editorial with the same headline on 9 July 2012, which said:

[T]he Karnataka government has been scrambling to put together plausible arguments for rejecting this much sought-after tag for atleast ten of these pristine places which fall in the state. The government has been steadfastly opposing the tag in the run-up to the announcement, but UNESCO has ignored it and gone ahead with the declaration . . . the government is trying hard to play the tribal displacement card, arguing that the heritage status would result in eviction of local tribals in order to preserve the flora and fauna . . . the government has not only failed to preserve the Western Ghats but allowed its slow deterioration. It should realise the UNESCO tag brings with it the urgent need for conservation of not just these sites but other too as suggested by environmentalists, and mobilize resources to make sure it's done.

The strong editorial stance taken by both the dailies could not stand the test of time, as it was confined to July 2012 and even though the matter has not been resolved, with the government yet to take a stand on ESAs, the media seems to have lost interest and purpose in this crucial matter.

On comparing the two national dailies, we found that the results show that there were clear and significant ($X^2 = 172.176$; df = 5; $p < 0.000$) trends (see Figure 4.10). *Deccan Herald* took the lead in reporting largely about issues pertaining to the controversial demarcation of the ESA, while *TOI* turned toward sustainability and long-term solutions in mitigating Climate Change in its focus of coverage.

The study identified relationships between arguments and policy-related issues, including definition of the problem; appropriate solutions; political support; technical feasibility and accountability for implementation through these frames:

1   Related to ESA reports
2   Related to conservation/wildlife
3   Related to tourism/development
4   Related to sustainability
5   Related to forest community
6   Related to energy/resources/hydel projects

A majority of arguments in the news reports were focused on the policy problem itself. Within that category, arguments were predominantly about

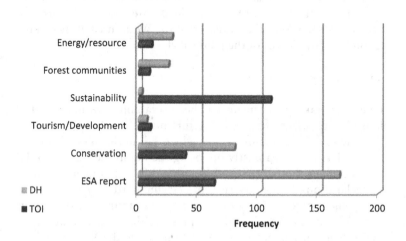

Figure 4.10 Classification of issues focused on the Western Ghats in *Deccan Herald* and *TOI*

ESA report (72% of 552) and sustainability (97% of 552). Political arguments predominated on the policy issue (230 of 552), followed by conservation and sustainability (42%). We found very few arguments (34 claims) related to forest communities. The near absence of arguments related to that issue might be explained by the fact that the policy was new and that implementation as well its monitoring and enforcement had not been initiated during the period of the study. It might also reflect a relative lack of interest in these matters on the part of the news media, as these forest communities do not constitute their "readership" and "advertisement" base.

The most interesting finding through this media content analysis, we know that on balance the news reporting is overwhelmingly supportive of the goals and principles, as noted in the study related to the Montreal Protocol (Howland et al., 2006). More than half of all the arguments were about the problem. Examining the data over time, we see that spikes in news coverage – and by extension, spikes in rhetorical arguments – correspond with big news events such as the heritage status to the Western Ghats, international policy conferences, reports of scientific expedition results and political controversies. A review of these arguments enables us to assemble a timeline of what happened.

McComas explores whether "narrative factors explain change in media coverage of global warming over time" (McComas and Shanahan, 1999, p. 38). The study of 312 articles, written between 1980 and 1985 in *The New York Times* and *The Washington Post*, records the presence of eight "themes." The study found, among other things, that during a phase of increasing news coverage, stories focused mainly on the consequences and implied danger of Climate Change. "From a narrative standpoint," wrote the authors, "news coverage in the late 1980s had set up an atmosphere in which global warming was an imminent disaster" (McComas and Shanahan, 1999, p. 52). The discussion considers the implication that such prediction could discourage media attention to the issue. This approach is useful for gauging the tenor of news coverage surrounding a policy problem. A study by Wilkins (1993) analyzed US news stories about Climate Change from 1987 to 1989 to discern values that help frame news about the greenhouse effect. The author billed this study as a "qualitative analysis with (quantitative supporting information)" (Wilkins, 1993, p. 75). A wide array of information was collected: coding categories included the media outlet, the month and year the story was produced, whether the stories were news, feature or opinion, the number and type of sources cited, the news peg, the amount of coverage devoted to the greenhouse in each story, the metaphors used, whether the story discussed the future and how politics was treated in each piece (Wilkins, 1993, p. 75). In all, 1441 articles from *The New York Times*, *The LA Times*, *The Washington Post*, *Associated Press* and *Time* were coded with intercoder reliability measured at 0.85 using a straight percent measure. The study concluded that three yet unexplained values help frame news coverage of Climate Change: progress, the institutionalization of knowledge and innocence. The sample size and scope of information collected for this study by human coders are considerable. In addition to enabling an analysis of themes, this approach seems to have potential for describing social process within the coverage, for example, by comparing actor data with news pegs and metaphors. In a study of news coverage of Climate Change from 1986 to 1995, Zehr (2000, p. 85) reports that the news media used the theme of scientific uncertainty in Climate Change reports "to create an exclusionary boundary between 'the public' and climate change scientists" which "delegitimated lay knowledge." Unlike the previous studies, the methodology here is not thoroughly explained. Codes are not listed and no measures of reliability are reported (Howland et al., 2006).

On further analysis, the results show that there were clear and significant (X^2 = 4.521; df = 1; p < .033) insights (see Figure 4.11) in the manner in which the two dailies highlighted the problems around the declaration of Western Ghats as an Ecologically Sensitive Area through the lens of these stakeholders. Of the total sample size, *Deccan Herald* highlighted the plight of the forest communities (53%) through 155 articles, while *TOI* had 143 articles that brought out the voices of this marginalized community. Gram Sabhas (the primary body of the local village elected council, the Panchayat System in India) mandated by the landmark Forest Rights Act, 2006 (FRA), are the officially recognized space for these communities to participate in (Choudhury, 2016), and journalists must venture into the hinterland to bring out more of these local voices.

In terms of the rhetorical devices used by journalists to communicate issues pertaining to Climate Change, the results from our study show that

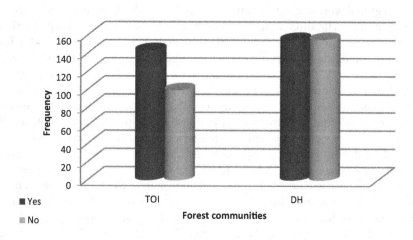

Figure 4.11 Comparison of coverage in *Deccan Herald* and *TOI* in highlighting issues faced by forest communities in the Western Ghats

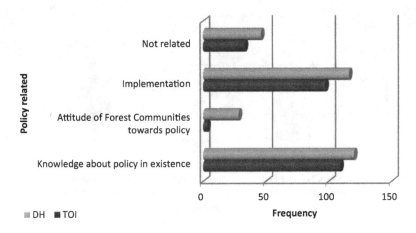

Figure 4.12 How do *Deccan Herald* and *TOI* journalists communicate Climate Change issues?

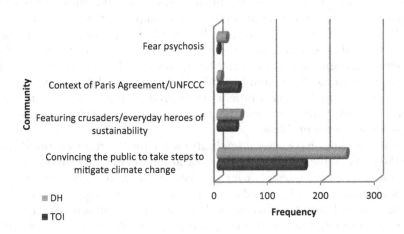

Figure 4.13 Context of communicating by journalists (rhetorical devices)

there were clear and significant ($X^2 = 16.396$; df = 3; $p < 0.001$) trends (see Figure 4.12). Most of the journalists communicated about the Ecologically Sensitive Area policy to a readership that was already aware of the implications of this policy. About 229 (41%) of the total articles reflected this journalistic device and that has been used as a frame to communicate this policy-related aspect in this sample of the articles we collected.

When looking at the context through which journalists reported about Climate Change, the results show that there were clear and significant ($X^2 = 41.230$; df = 3; $p < 0.000$) trends (see Figure 4.13). Convincing the

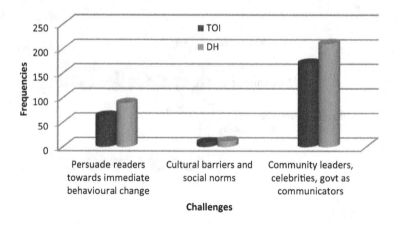

Figure 4.14 Challenges in communicating Climate Change and emergence of frames

public to take steps to mitigate Climate Change was the most widely used rhetorical device by journalists in this study sample that we analyzed.

When reporting on this topic, the biggest challenge journalists faced (see Figure 4.14) was restricting the news report to referencing sources from celebrities, community leaders and official sources. There was considerable scope here, to go "above and beyond" the call of their duty and investigate by scratching the surface a little more to produce impactful, yet unbiased reports – that truly portrayed the ground reality and explore the various dimensions of the problem.

These findings have several implications for further research and theory building. First, the results show that official sources dominate in the media's narrative of environmental coverage. Second, the results demonstrate that official sources are a large and heterogeneous category, suggesting that it is time to move beyond a focus on the dominance of official sources per se and toward research and theory that explores differences within the category of official sources and their antecedents. Third, journalists faced multiple challenges from official sources in communicating Climate Change. Fourth, convincing the public to take steps to mitigate Climate Change was the most widely used rhetorical device by journalists in this study.

Directions for media coverage

We do not know the trends in regional language newspapers, radio and television coverage, though they enjoy a much wider coverage. However, anecdotal evidence suggests that there is significantly less coverage of Climate Change in regional/vernacular Press. The study has not looked at other media, such as the radio and television, which have a wider coverage and can transcend the literacy barriers in a country like India. The correlation between media reports and readers can be explored to understand the significance of media reporting and its perceived outcome; the same framework can be extended to policy makers or elected representatives to seek their responses to media reports related to Climate Change.

The framing analysis of the coverage allows for conclusions about the coverage, but using the analysis to get to the effectiveness of the policy remains unclear. Some limitations would need to be acknowledged here, of the Press and its reach, of cultural change and its starting points. Further research can include media such as blogs and online resources of institutes and think tanks working in this area. A correlation between news articles and journalists who reported these articles related to environment and Climate Change can reveal insights into editorial interference and constraints faced by the journalists from their organizations. Studies have shown that journalists are unwilling to participate (Brüggemann and Engesser, 2017) as it is a challenge for them to search for new ways to frame Climate Change. Journalists and media scholars need to look at new dimensions in this debate and seek emerging narrative threads. The findings also emphasize the importance for shift and re-drafting of editorial policies within the broader Press production and content generation purview.

Increasingly, we are faced with evidence every day that Climate Change is a social and economic concern with direct effects on human capital in developing countries and emerging economies. Changes to the environment in the form of disasters or, as in this case, earmarking a hitherto common zone into an ESA can wreak havoc in existing relationships between man and nature. This can result in disparities in income and political voting behavior too. Political polarization and influence of voters based on this concern can throw up extreme and imbalanced forms of political systems. Ultimately, this leads to ineffective policy instruments that can be detrimental to productivity and increase migration toward urban clusters resulting in immense pressure on resources.

Note

1 This could include a news release from this organization.

5

CLIMATE-COMMUNICABILITY

Saving our planet, lifting people out of poverty, advancing economic growth . . . these are one and the same fight. We must connect the dots between climate change, water scarcity, energy shortages, global health, food security and women's empowerment.

~ Ban ki-Moon, former UN Secretary-General[1]

Many different media are used by audience members in today's mediatized world (McQuail, 2005, p. 455). Media are means of communication that distribute content – such as text, pictures and sound – to an anonymous and spatially diverse public via technical means (McQuail, 2005). Print media includes printed publications such as newspapers, magazines, books; broadcast media comprises radio, television or films; digital media includes websites, social media platforms and media used by stakeholders such as non-governmental organizations, political parties, or companies.

Media representations of Climate Change have inevitably become the main source of information for citizens and decision-makers alike. Social media has also been used for social movements and digital activism by climate crusaders to create a shared identity and collective goal. These climate crusaders, specifically choose to time their activism when major summits are organized around the globe, where there is a congregation of world leaders and scientists. Sensing a need to reach out to this segment of the audience who consume news through Twitter and other social networks, the organizers of COP 23 created hashtags, visual content and dashboards specifically around the summit to monitor tweets and measure virality (sum of likes and retweets can measure virality of a tweet. When a user "favorites" or "likes" a tweet, it displays they approve the tweet).

Early media coverage primarily focused on links between weather, food and climate and demonstrated the deep roots of links that have been made between them. According to Boykoff (2011), during particularly cold

DOI: 10.4324/9781003015673-5

winters in the Northern Hemisphere, the *Edinburg Advertiser* ran a story in September 1784 that read:

The year 1782 proved remarkably wet and cold, the crops over a great part of Europe were more or less injured, and the northern climates experienced a ferocity, amounting to a famine. . . . Potatoes, which in bad seasons had proved a substitute for grain, were this year frost bitten, and rendered entirely useless.

(p. 41)

Between 2003 and 2006, there was a steady increase in attention on Climate Change among the opinion-leading newspapers in the United States, with an even more marked rise in the United Kingdom (Boykoff, 2007; Boykoff and Roberts, 2007). The release of Al Gore's film *An Inconvenient Truth* clearly played a part in generating the rise in media coverage of Climate Change in 2006. Over this period, there was also an increase in coverage in Australia/ New Zealand, the Middle East, Asia, Eastern Europe and South Africa, though this was less steep than in the UK (Boykoff and Roberts, 2007, p. 39). Ironically, there has been comparatively little media coverage of Climate Change in developing countries; yet they are likely to suffer the worst effects. In 2006, there was four times as much coverage of Climate Change issues in the UK Prestige Press than there was in 2003, whereas in the United States coverage increased in the elite National Press by roughly two and a half times for the same period (Anderson, 2009; Boykoff and Boykoff, 2007).

"Climate change is a notoriously difficult subject for journalists to report on, for editors to maintain interest in, and for audiences to grasp" (Painter, 2010, p. 3). Moreover, journalists, like scientists, look for objectivity and fairness in reporting. Hence, an interesting correlation exists between the increase in the number of peer-reviewed scientific papers that are being published and the increase in these research papers being cited as evidence in media reports by journalists. Research shows that the promotion or presence of uncertainty in media reporting of climate science can act as an obstacle to public understanding and lead to disengagement, so it is critically important that newspapers provide proper context when covering such an important issue (Happer and Philo, 2013).

In the process of understanding media representations of Climate Change, what is the understanding about this global issue among the people likely to be most vulnerable and marginalized? Is there any awareness about Climate Change? Is that awareness driven by key informants, such as a field supervisor working with a not-for-profit in that area? If there is awareness, where does it come from? Has the media played a pivotal role in shaping this understanding?

Brainard (2008) wrote in the *Columbia Journal Review*:

> Short-term changes in the [weather] are irrelevant to the long-term trends in the [climate]. Yet every winter, the onset of cold inspires climate skeptics to once again attempt to "debunk" global warming and journalists to once again fall for the maneuver . . . despite journalists' earnest and somewhat successful efforts . . . lingering confusion about the basics of climate science continues to plague public understanding.

Media attention

Why does media coverage matter? What is missing from media coverage of Climate Change? Media reporting has focused on political developments, natural and man-made disasters and socio-economic development. However, does the media speak for the climate (Boykoff, 2011; Bushell, Buisson, Workman, and Colley, 2017) and for those marginalized populations affected by Climate Change?

The concern over carbon emissions, global warming and Climate Change has been in the news over the last decade. The media's portrayal of Climate Change has attracted the attention of scholars and researchers from interdisciplinary areas such as economics, environmental studies, communication and media, as well as pure sciences. The interdisciplinary nature of this study is warranted given the significance and scale of the issue. Climate Change is not confined to a single nation or community. It is an issue that calls for borders to be blurred and governments to unite to make policy decisions and implement them. Though studies on the scientific implication and mitigations of Climate Change have been of interest to researchers globally, the understanding of the implications of this research needs to be conveyed to policy makers, and this is where the media acts as a channel to communicate the understanding of this science in simple terms.

Asian countries in particular face risks, impacts and the need to mitigate Climate Change given their growing economies and population. It becomes important that the media reflects the importance of this issue, which affects public opinion and policy.

COP summits

Studies have shown that annual UN climate summits receive intense global media coverage. This coverage not only keeps the local public engaged and leads to an interesting exchange of knowledge about climate politics but also mobilizes people to combat Climate Change. A study conducted on German public's exposure to news about the 2015 Paris Summit found that people did not engage with the discussions in a more active way and were less likely to support the government's policy decisions on the issue (Brügge-mann et al., 2017). This shows a disconnect between the media's portrayal of the issue and the public's understanding of it.

Another study of US media coverage of the 23rd session of the "Confer-ence of the Parties" (COP 23) found that Climate Change stories lead into wider considerations of attention paid to political content of coverage.

In November, when the COP 23 summit was held at Bonn, media attention to climate change and global warming was up just slightly (3%) throughout the world from the previous month of October. Increases were detected most strongly in Asia, the Middle East, Africa and Europe, with a decrease in Oceania and counts holding steady in the Americas. Compared to counts from fifty-two sources across twenty-eight countries in seven regions around the world in November 2016, the global numbers were actually down about 23%. The high levels of coverage in November 2016 were largely attributed to the US Presidential election of Donald J. Trump and the concatenate Marrakech round of international climate negotia-tions (COP 22). While in November 2017 was punctuated with the Bonn round of climate talks (COP 23), it did not prove to be nearly as resonant a media event-come-story as those that unfolded in the previous November. At the country level, coverage was also up from the previous month of October in Germany (17%), India (21%), Spain (8%), the UK (14%), and the US (2%). Coverage was down in Australia (–39%), Canada (–7%) and New Zealand (–29%).

(Boykoff and Boykoff, 2007)

COP 23 in Bonn

COP 23, under the UNFCCC, which aims to "prevent dangerous anthropogenic interference with the climate system," took place in Bonn, Germany, from 6 to 17 November 2017. The summit was significant as the US government was not part of the international climate agreement. In this arena, the November 6–17 international climate talks (COP 23) in Bonn, Germany, dominated news cov-erage, even as the tiny city on the banks of the River Rhine welcomed 23,000 delegates to negotiate the rules of the Paris Agreement signed two years earlier.

As pointed out earlier, Climate Change is a complex issue and receives unprecedented attention from world leaders, the media, not-for-profits, activists, industries and common citizens. It was in November 2013, at COP 19, the 19th session of the COP to the 1992 UNFCCC, and the ninth session of the Meeting of the Parties (CMP) to the 1997 Kyoto Protocol (the protocol having been developed under the UNF-CCC's charter) at Copenhagen, that media reporting related to Climate Change surged.

It is noteworthy that the convergence of world leaders and influencers brings a global environmental issue such as Climate Change to the center stage. This was evident at COP 19 held in Copenhagen, where nearly 130 heads of state from 191 countries and over 40,000 scientists, activists and industrialists registered for the convention.

The purpose of this study is to contribute to the growing literature by offering an assessment of Twitter using UNFCCC's digital tracking platform to get an overview of its engagement during the COP 23 Bonn conference held in November 2017. Twitter has become an influential platform to be seen and heard. Politicians and influential actors, such as activists, celebrities and sportspersons, known as the elite actors on Twitter, have a strong presence and seek to share information and their ideologies through their tweets. Scholars refer to this phenomenon as the need to seek gratification while using the medium actively (Blumler and Katz, 1974; Coleman and Blumler, 2009; Ganapathy, 2015).

A content analysis of tweets by Indian politicians and celebrities in India showed that a majority of the politicians (79.3%) tweeted to spread awareness and indicate their position on social issues (73.3%) while celebrities' tweets did not spread awareness about issues. A majority of the celebrities (82.9%) did not reveal their position on social issues nor did they convey their personal ideology or influential role in their tweets (Ganapathy, 2015). Hence, this highlights the need for celebrities to express their concern for issues relating to the environment and social causes in greater detail so as to influence their follower base. An event like COP 23 can be used by politicians and celebrities to drive discussions and focus on this significant issue of our times.

While social media is a term that encompasses platforms as diverse as online social networks, blogs, microblogs and video- and photo-sharing sites, it is services such as Facebook and Twitter that have garnered the most attention for their use in political discussion, engagement and mobilization, especially among the young. Moreover, social media is viewed as a potentially effective means of improving the relationship between citizens and their representatives (Coleman and Blumler, 2009). However, media stories on Climate Change or global warming typically manifest through primary yet often intersecting *political, scientific, cultural* and *ecological/ meteorological* themes (Boykoff et al., 2018).

In such a scenario, the analysis of tweets around an important environmental summit can reveal insights into Twitter usage and message content creation. Twitter lends itself to being an effective and strategic platform that is high on visibility. Its user base consists of influential and key policy makers who are involved in climate policy decisions. Since considerable amount of diplomacy also takes place between Heads of State and Embassies, it is considered a platform very high on "being seen" and "tweeting facts" especially if the conference organizers use hashtags and mention keywords to ensure the conference is "trending" for the given period of conference days.

Since digital media platforms such as Twitter generate volumes of big data, they allow us to examine communication efforts in greater detail and provide interesting insights into strategies to improve the content and reach. Researchers have found that digital technologies like social media and big data not only reinforce and expand public diplomacy discourses but also become a driving force in change management in the structure and process of diplomatic work (Zaharna, Arsenault, and Fisher, 2014; Bjola and Holmes, 2015).

Climate Change communication is a diplomatic exercise involving more than 150 nations and organizations. In the COP 23 summit, the role of digital media as an advocate and channel for diplomacy was amplified.

Figure 5.1 This word cloud shows the frequency of words (four letters or more) invoked in media coverage of Climate Change or global warming in *Los Angeles Times, The New York Times, USA Today, The Washington Post* and *The Wall Street Journal* in the United States in 2017

Source: http://sciencepolicy.colorado

Conferences as massive in scale as this Global Environment Summit have dedicated social media measurement matrices and analytical tools to map and measure engagement and drive discussions. Special search parameters such as hashtags and key words are used to pool in all the data. In the COP 23 conference, #COP 23 and #climatechange were the two hashtags used to generate 893,450 tweets.

In Figure 5.1, the coverage of COP 23 in US media is depicted through a word cloud. The absence of US government from COP 23 is particularly covered by the US media, as is evident in the word cloud generated.

A study carried out on the COP 21 coverage demonstrates some differences between some of the recently emerged "digital-born" players and legacy media and between the new players themselves. The media outlets in this study were driven by their desire to be different from legacy media and to publish content that would be widely shared by younger audiences on social media platforms (Painter, Kristiansen, and Schäfer, 2018, p. 8).

When a summit such as COP 23 is organized, it presents an opportunity for redirecting online information flow. The release of an IPCC report, presented by a range of diverse and decentralized actors, showed the indirect communication flows on Twitter presenting opportunities in which elites are more likely to have their message rebroadcasted to primary and secondary audiences, thus expanding their reach and scope (Newman, 2016, p. 9).

The COP 23 summit began on 6 November 2017, but the time range was set from 5 to 17 November 2017. Previous studies focusing on this type of research found it necessary to be careful about setting the time range of the search parameters. It is not sufficient to collect data on the day of the campaign; data must be collected on the days before and after to track possible changing trends. Take the examples of studying the Arab Spring movement and the People's Climate March. The former was a social movement with a longer duration, and thus researchers examined the Twitter data for several months to capture the big picture; while in the case of the People's Climate March, a one-day event in New York City, it was sufficient that researchers only looked at the tweets the day before, on the day and at the day after for analysis. In a word, the time range we choose for search parameters should be consistent with our research goal (Social Media Analytics for Digital Advocacy Campaigns: Five Common Challenges, 2016, p. 5).

Since this study relied on the official UNFCCC social media measurement tool – www.climatetalkslive.org – which was outsourced to KPMG (a consulting firm)[2] for the purpose of this study, the hashtags #COP 23, #climatechange and #climate were used. This resulted in a total of 893,450 tweets. On the first day of COP 23, the number of tweets generated were 32,453. It is also interesting to note that in a matter of 24 hours, the number of tweets surged to 220,582. On the first day of COP 23, Germany generated

61% of all tweets. This was because Bonn, the venue for COP 23, is in Germany. The sentiment analysis on the first day showed that 73% of all tweets were negative, and 53% of all tweets were generated by males. This gender segregation of tweets points to the role of elite actors on Twitter and their dominance in the discussions.

Another interesting phenomenon was the surge in tweets on 08 November, 2017, especially from the US region. This shows that government leaders were more likely to give the discussion prominence, both before and during their speeches at the summit, tweeting actively about it; however, the interest died down after they left the summit.

By the middle of the conference, the sentiment analysis showed that the number of negative tweets had drastically reduced, and 88% of tweets were found to be positive. The United States was leading (20%) the conversation surrounding President Trump's withdrawal from COP 23, while European nations were lagging behind in discussing about the summit on Twitter. This indicates that US media publications were cued into the happenings at COP 23 and were reporting about the talks while allocating substantial space in their coverage.

The positive tweets generated related to sustainable solutions and Climate Action stories as well as stories from Indigenous communities, especially communities in island nations such as Fiji and Kiribati. These stories of positivity and resilience were shared and retweeted, resulting in considerable "sharing of voices" on Twitter.

The analytics from the UNFCCC's official measurement website, www.climatetalkslive.org, had segregated the elite actors on Twitter under "Participants" profile. These included media accounts, journalists, influential environmental activists, communities working for the environment, government leaders and officials of the United Nations. The analytics here showed how each tweet, retweet or reply by these elite actors on Twitter, in turn, reached their millions of followers.

Apart from this analytical tool, the UNFCCC website had details about Live Twitter chats, Facebook chats and other Digital Surge workshops. For example, presented here is one of the ways to reach out to audience on digital media platforms:

> Join us Monday, November 6 as we kick-off COP 23 with the #Uniting4Climate Global Digital Surge, hosted by Connect4Climate, United Nations Foundation and the Fijian COP 23 Presidency. To advance climate action further, faster, together, the Surge will showcase how an unstoppable Grand Coalition of governments at every level, civil society, businesses, universities and individuals are all "Uniting for Climate Action." Use the hashtag #Uniting4Climate to join this worldwide conversation and share your bold climate solutions.

The UNFCCC social media team did a commendable job in understanding the needs of its various stakeholders, internally and externally, to craft a communications campaign around #COP 23. The reach and impact of this continuous and relentless communications campaign were picked up by the media and elite actors on Twitter to broadcast to a wider audience.

However, mainstream media can go beyond the official sources available to craft stories that have a deeper impact on society. The media cover Climate Change under three circumstances – major events like COP 23, disasters such as tsunamis and hurricanes and when world leaders discuss policy-relevant matters such as the Kyoto Protocol and Paris Agreement.

The role of journalists should not be limited to the coverage of Climate Change conversations during risk or uncertain conditions. Climate Change is a pressing issue of our times, and it warrants equal action from citizens, governments and the media. Communicating this to the government and citizens and bridging the divide is a critical role that the media has to take cognizance of.

If Climate Change and mitigation reporting witnesses a surge only during events like COP 23 and dies a slow death after the buzz surrounding an event is over, it will be a challenge to consistently build consensus to keep the pressure on world leaders to contemplate and formulate policy changes to save planet Earth.

Notes

1 Quoted in Jones, R. (2017). *Confronting complexity* (p. 302). London, UK: The X-Press.
2 https://home.kpmg/xx/en/home/about/who-we-are.html

6

CREATING COMMUNITIES TO REPORT ON CLIMATE CHANGE

Around 1.6 billion people depend on forests for their livelihood. This includes some 70 million indigenous people. Forests are home to more than 80% of all terrestrial species of animals, plants and insects. In India, the Chipko (tree-hugging) Movement is the most widely known example of environmental leadership to demonstrate that Climate Change impacts indigenous landscapes and marginal groups are increasingly susceptible to climate instability and continue to bear a disproportionate burden to adapt to Climate Change.

The UNFCCC refers to Climate Change adaptation as "changes in processes, practices, and structures to moderate potential damages or to benefit from opportunities associated with climate change." IPCC elaborates adaptation as adjustment in natural or human systems in response to actual or expected climatic stimuli or their effects, which moderates harm or exploits beneficial opportunities (IPCC, 2014a, 2014b).

India's starting point is the urgent need for poverty alleviation and more energy for development. In 2016, 21.9% of Indians lived below the national poverty line, while more than 240 million people lacked access to electricity (ADB, 2017; IEA, 2016). The country, in line with many others, faces a dilemma in addressing its developmental goals: it needs to respond to demands in poverty reduction, energy access and urbanization while reconstructing its development pathway toward a cleaner energy system that has been coupled historically with fossil fuel use. Coal has remained the mainstay of India's electricity generation, contributing to 61% of the total national generation capacity. With the goal of providing power more efficiently to more people, India is investing in modernizing its power plant fleet.

Since 2006, the country has added 151 GW of new coal power, making a total of 218 GW as of June 2017, with about 75% of this capacity being subcritical.

As such, afforestation and reforestation are considered established CO_2 removal options, and projects could feasibly be launched soon. These uncertainties could be reduced or better characterized through a dedicated science program aimed at understanding the issues across ecosystems, latitudes and climate zones. There is also the added potential to contribute to other global

DOI: 10.4324/9781003015673-6

sustainability goals such as improved water quality, ecosystem restoration, biodiversity preservation and job creation.

Case study – narratives of forest communities

Since we started rebroadcasting the COBAM radio programme [on climate change and forests in the Congo Basin] people – particularly farmers – have been coming to our studio for more information on how to deal with unpredictable seasons and have better harvests.

~ *Mngo Demse*
Community radio journalist in Bamenda, Cameroon
(UNESCO, 2013)

"The climate is changing." The traditional knowledge systems of the many native tribes inhabiting the forests of the Western Ghats may no longer be able to predict the symbiotic relationship that existed between nature and the tribal communities for many centuries.

Global Climate Change is likely to impact the cultural, spiritual and often economic ties that inextricably link the tribes such as Soligas, Jenu Kurubas, Kurubas, Todas, Badagas and many other communities who lived inside the forests and have now been forced to live on the periphery. The traditions of many native cultures (Alaskan Natives, Native Americans, Maoris, Aborigines, etc.), such as indigenous foodways, hunting practices and plant gathering, are threatened. In regions where the impacts of Climate Change are already pronounced, traditional knowledge systems can no longer predict the cycles of the planet that have provided harmony and balance between nature and the needs of tribal communities for eons. Forest communities, especially in the Western Ghats, have an ecological, spiritual and food security connection with the forest they inhabit. A few examples in this region are the sacred groves in Kodagu district in Karnataka and the sacred groves of the Soligas – Sankamaboodi, Paadadhare, Karayanabolli, Byadanaborebolli, Tavasaarekaadu in the MM Hills region. This earmarking of boundaries for uninhibited forest growth are firm concrete efforts taken by these forest communities who revere and worship nature toward afforestation (Bhagwat, Kushalappa, Williams, and Brown, 2005).

To gain a better understanding and knowledge of how forest communities live in harmony with nature, the researcher visited the hamlets of indigenous tribes living in the region surrounding MM Hills in Chamarajanagar district of Karnataka. The region was specifically chosen as the tribal communities are recovering from the aftermath of the dreaded poacher Veerappan's clutches over their villages, as well as cops of the Special Task Force who were stationed in the area. Hence, the exposure that the tribal communities have to development is still in a nascent stage as compared to other communities in the Western Ghats.

The purpose of going to the field and interacting with these communities was to understand the impact of media reporting on Climate Change on communities that were affected the most by the changes to the environment as well as to document the effects of policies concerning conservation and development on these communities' sustenance and livelihood.

Why is this important? Doom-laden depictions of Climate Change are ubiquitous in the media. Splashing the front page of newspapers with images of flooding occurring in big cities, relentless coverage of heavy rains and loss to lives and property on television and radio can create a tremendous impact in the manner in which viewers and readers process and perceive these horrific visuals. However, there is a lack of clarity in the literature about the impact that fearful messages in Climate Change communication can have on people's senses of engagement with the issue and associated implications for public engagement strategies. Some literature suggests that using fearful representations of Climate Change may be counterproductive. The authors O'Neill and Nicholson-Cole (2009) explore this assertion in the context of two empirical studies that investigated the role of visual and iconic representations of Climate Change for public engagement, respectively. Results demonstrate that although such representations have much potential for attracting people's attention to Climate Change, fear is generally an ineffective tool for motivating genuine personal engagement. Non-threatening imagery and icons that link to individuals' everyday emotions and concerns in the context of this macro-environmental issue tend to be the most engaging. Recommendations for constructively engaging individuals with Climate Change is the need of the hour, to suit different sections of the society, with a clear understanding of their receiving knowledge capabilities in the communication process. Through the emerging conversations during the course of the FGD, it was apparent that the disaster narratives gained attention, but were not enabling enough to motivate the participants to engage meaningfully and motivate themselves as well as their community to adopt a behavioral change.

Theoretical framework

Behavioral change communication, which is a part of the development communication, specifies how communication influences intention and behavior and has its origin in behavior change theories. There are other models and theories which play an important role in developing behavior change communication. The theories include diffusion of innovation model by Everett Rogers, stages of change model by Prochaska, DiClemente and Norcross, self-efficacy model by Bandura, behaviour change continuum by World Bank. The process of behavioral change communication involves the use of mass media and community mobilization to bring about the desired change.

Keeping in mind this theoretical framework, eight focus group discussions (FGDs) were conducted in five villages of MM Hills region (see Table 6.0).

Table 6.0 Details of villages where FGDs were conducted

Elevation (3000 feet)	Number of households	Distance to main road (minutes)	Number of FGD participants
INDIA			
State: Karnataka			
District: Chamarajanagar			
Village: Dorsane Grama	55	10	8
Komdikki Grama	30	45	15
Palar Grama	50	30	15
Anehole	28	5	9
Mahadeshwara Betta	20	20	14

A purposive sampling approach was undertaken through selection of interviewees relevant to the processes as well as the focus of the study. The aim was not to provide a sample that represented all those forest communities that reside in the Western Ghats in Karnataka but to particularly amplify the voices of those forest dwellers who were the most affected by the declaration made by UNESCO and the subsequent policy decisions. Our respondents were individuals who were approached because of their involvement in some form or the other with relation to their awareness and activism within the constitution of forest rights committees. The interviewees were contacted through the extensive network of contacts of the author of this study. The methods applied were mainly qualitative and consisted of FGDs, semi-structured interviews at the household level and participant observation. The fieldwork was conducted between September 2017 and October 2017. Eight FGDs were held in each of the five villages, comprising a mixed group of men and women (aged 25–55), and 65 in-depth interviews were carried out. Participants were selected based on their availability (given the fact that they are daily wage laborers and time constraint of the project). Ten senior citizens (aged 60 and above) were particularly included in the FGD (15% of the sample) as they bring with them insights on the "change in climate" over the years.

The study was built on the assumptions that

(a) climate variability and change are noticeable to the tribes and directly affecting their livelihoods;
(b) climate variability is not new to people who live in close proximity with nature, and they have a wealth of traditional knowledge and response systems to cope with;
(c) key informants among these tribes create awareness and inform them about media reports relating to Climate Change and of any immediate impact;

(d) exposure to media (radio, television, newspaper) narratives about Climate Change leads to better understanding of the effects of Climate Change.

The following research questions guided the study

1 How do tribes living on the periphery of the forests of the Western Ghats perceive and interpret Climate Change?
2 What are the major impacts of the Forest Rights Acts on the communities' livelihoods?
3 Has media exposure through representations about Climate Change communicated visible behavioral change?

The data were gathered from eight two-hour FGDs facilitated by the researcher, involving 60 volunteers. Following the FGDs, given the exploratory nature of the research and the fact that it was based on volunteers, the reasons for the low female participation are hard to ascertain, but this also suggests that there may have been gender-specific attitudes toward these topics. For instance, the men in the household may be seen as the authority to talk about these matters whenever called upon. This can be an interesting area for further research. We would like to stress here that the following findings are based on an exploratory study, so this data cannot be considered to be generalized.

Studies on the scientific implication and mitigations of Climate Change have been conducted globally. However, the risks, impacts and adaptation to Climate Change need more in-depth research – especially from developing countries. Asian countries in particular are facing the problem on such a massive scale that efforts may not be adequate due to lack of resources and low priority. This may delay the developing countries' readiness to Climate Change impacts.

Policy makers should review their development plans through the climate lens and identify the linkages between climate and development. The Asian region is found to be more vulnerable to Climate Change compared to other regions due to its large population and its large dependence on natural resources and agriculture.

The most striking finding in both interviews and FGD responses was individual's confusion over the very concept of Climate Change. While many found the concept "interesting" or "challenging," at the same time, it was generally considered a slippery topic to deal with in their day-to-day existence.

Communities' perceptions of Climate Change

Marginalized groups often live, in high-risk areas, have lower coping capacities and have limited or no risk cover in the form of insurance or other safety nets. Further, they are heavily dependent on climate-sensitive primary industries – notably agriculture, forestry and fisheries. Thus they risk both injury/death and major disruption to their livelihoods. Other livelihood activities of marginal groups such as labor for construction, selling fruits and vegetables, etc., belong to the informal economy, making it difficult to assess disaster impact on them through frameworks and indicators designed for formal economic sectors (Smith et al., 2014; Vachani and Usmani, 2014).

Lately, economists and conservationists have begun touting community-based forest management (CFM) as a way to alleviate poverty as well. They frequently offer the approach as a means to achieving the United Nations' Sustainable Development Goals of ending poverty and hunger and reducing ecological degradation by 2030. CFM's role has also been highlighted in talks about the UN's global initiative to mitigate Climate Change known as Reducing Emissions from Deforestation and Forest Degradation (REDD+). The belief is that REDD+ and CFM overlap substantially in their goals of achieving forest protection and generating socioeconomic benefits for forest-dependent people.

Climate Change and its association with human trafficking have been studied in the context of natural disasters, crop failure and socioeconomic conditions in developing countries.

Puerto Rico is struggling to come to terms with the loss to life and property after Hurricane Irma destroyed its economy recently, even though it's a province of the US, the most developed country in the world. Hence, the effects of Climate Change in developing countries like India can be disastrous.

"Havamana badlavane" (Climate Change in Kannada) was the term used during the FGD (see Table 6.1). Elders in the FGD highlighted the fact that precipitation has decreased due to which they face lot of financial loss when monsoons fail.

> [B]ut we do not know the reasons why this happens. We know that we have to face a lot of hardship; we have wells for water, now the state government (4 years ago) brought water tanks to our villages. But we are aware of the fact that unless there is ground water, there will be enough water in their wells- and for the ground water to recharge, we need adequate rain.

Asian monsoon, one of the largest climate systems in the world, existed 40 million years ago, much earlier than previous estimates, according to a new study. Scientists first believed the Asian monsoon began 22–25 million years ago as a result of the uplift of the Tibetan Plateau and the Himalaya Mountains. A research team has now found that it existed 40 million years ago during a period of high atmospheric CO_2 and warmer temperatures.

Macchi, Gurung and Hoermann (2017) highlight that Himalayan populations living in the vulnerable mountain areas have decades-old indigenous

Table 6.1 Overview of community responses to climate and socioeconomic change

Variables	Perceived change	Response of the community
Temperature	Prolonged hot spells	Health problems
Precipitation	Erratic, sudden cloudburst	Changes to crop varieties, labor migration, crop failure, skipping meals, sale of assets, borrowing money
Onset of monsoon	Delayed, unpredictable	Watershed management, building tanks
Water availability	Dry springs	Less land cultivated, women and girls walk long distances to fetch water school dropouts
Pests and diseases	Mosquitoes, new pests	Malaria, dengue, chikangunya
Invasive species	Lanterna encroaching forests	Parasite/grows fast
Wildlife	Rare sightings of fox, elephants, leopards, birds	Man–animal conflict, children unable to differentiate animal sounds
Forest produce	Low produce – honey, dhoop, amla, Shikakai	Threat to livelihood, migration to coffee estates, quarries

knowledge and that such knowledge can supplement scientific climate analysis. They study people's perceptions of climate variability and change to understand existing response strategies in lieu of future climate prediction, particularly to understand how different social groups deal with their vulnerability and respond to disruptive environmental development. In doing so, they use local perception to climatic changes as a way to supplement and provide recommendations to the scientific weather data critical for researchers and policy makers. The study reflects that marginal populations' perceptions of change were highly consistent with the recorded climate data in the Himalayan regions.

A viable Climate Change policy strategy requires acknowledging the local and indigenous knowledge by developing appropriate adaptation strategies that are based on people's knowledge (Bose, 2017).

The FGD participants have been keenly observing the telltale signs of nature, as they live in close harmony and share a symbiotic relationship with the forests.

> *Earlier, the forests were very dense and thick foliage surrounded the ground of the forest. Today invasive species (lanterna) can be found deep inside the forests and they are spreading everywhere.*
>
> *We have forgotten the call of the foxes that were roaming in abundance a few years ago. Our children will never know how the sounds of the forests were a part of our life.*

These narratives are specifically designed to motivate media reports on Climate Action and also serve as a wake-up call to forest officials about the changing environmental concerns that could have serious repercussions for the forests. There, these narratives are "strategic," arguing for action and are effective enough in closing the action gap. However, the media had very little role to play in bridging the information gap, as these communities related media coverage to seeing visual images of Bengaluru city on their television sets coming to a standstill thanks to incessant rains.

Impact of Climate Change on livelihoods and community well-being

We are not allowed to go into the forests, for worship, we cannot take our children and perform puja. Heat is more, we do not know why. We are not supposed to go into forests and cut trees for firewood.

The Western Ghats, the biodiversity hotspot declared as a Natural World Heritage Site (NWHS, a formal designation through the United Nations, is globally recognized as containing some of the Earth's most valuable natural assets. Allan et al. (2017) in a recent study used two newly available globally consistent data sets that assess changes in human pressure (human footprint) and forest loss (Global Forest Watch) over time. These metrics were used to analyze spatial and temporal trends in human pressure for 94 NWHS and forest loss in 134 NWHS, presenting baseline data for the World Heritage Committee and the States Parties. India was subject to the highest levels of human pressure of any NWHS –the Manas Wildlife Sanctuary in India underwent the largest increase in human pressure of any NWHS, with its human footprint rising by 5 to a score of 17 and is now one of the most highly modified by humans (2017).

In India, a long struggle against such exclusionary forest policies and conservation practices resulted in the enactment of the Forest Rights Act (FRA), 2006. The FRA was initially conceived as legislation aimed at giving forest dwellers rights that they had been historically denied. The FRA also serves as a crucial barrier to one of the main drivers of biodiversity decline in India, namely the diversion of forests for developmental purposes (Lele, 2017). The FRA empowers forest dwelling communities to preserve their habitat from "any form of destructive practices affecting the cultural and natural heritage," and empowers them to "stop any activity which adversely affects the wild animals, forests and biodiversity."

These two powers have been further reinforced by a government circular (MoEF, 2009). The forests come under the State list. The distribution of power between Centre and States in India has certain environmental and

Climate Change topics coming under a list (concurrent) which means that both Centre and State have jurisdiction in these matters (Nayak, 2017). Vagrancy; nomadic and migratory tribes; prevention of cruelty to animals; forests; protection of wild animals and birds come under the concurrent list.

However, since the State has powers over the forests in its territorial boundaries, a major policy-related issue plaguing the Western Ghats since 2012 has been the debate over the implementation of two reports – one constituted by the State and the other by the Centre – Gadgil and Kasturirangan reports.

Caught in this crossfire are the tribal communities who depend on the resources in the forests for their livelihood.

> We used to go into the forests and pick products out of which we made brooms, and sold produce such as soapnut, dhoop (used in incense sticks), amla (Gooseberry), and honey. With the implementation of the Forest Rights Act, we have a limit to the number of products we can pick from the forests. Also we are regularly harassed by forest staff whenever we venture into the forests and false allegations are made.

Do they know about government policies related to environment? Not much, except that they are not supposed to go into forests and cut trees for firewood MoEF (2013).

Most of the participants of the FGDs are part of the Forest Rights Committee (1 committee=15 members). There are six committees under the Kollegal Soliga Samaja.

The interactions with the people who are most affected by Climate Change and policy decisions can rarely be attributed to a single driver. These findings should, therefore, be interpreted as the result of multiple drivers of change, related to rapidly changing climatic, socioeconomic and cultural conditions. The changes observed by the communities related mainly to food, water and income security; workloads and health (Macchi et al., 2017).

Sustainable methods in farming to grow ragi (Finger Millet), jawar (Sorghum or White Millet), avare (flat beans) are now being used by these communities. The main impact of change perceived by the communities was a significant reduction in outputs of staple crops, mainly due to reduced overall water availability and lack, or inappropriate timing, of precipitation and increased occurrence of crop pests and diseases, which the communities attributed to changes in weather patterns, increased use of fertilizers and monocropping. Prolonged dry seasons result in a drastic reduction in the availability of grass and other sources of fodder, as well as drinking water for livestock, forcing people, particularly women, to travel farther and longer to collect fodder, water and firewood. These findings were also found in the Himalayan ranges in Uttarakhand region and Nepal by researchers Macchi et al. (2017).

Community-based responses to policy decisions

Impacts of Climate Change on forests have severe implications for the people who depend on forest resources for their livelihoods. India is a mega-biodiversity country. With nearly 173,000 villages classified as forest villages, there is a large dependence of communities on forest resources in India. Chaturvedi et al. (2011) recommend that care should be taken to plant mixed species and planting should also be executed in such a way as to link the existing fragmented forests. Efforts should also be made to convert open forests to dense forests. Their analysis suggests that Western Ghats, though a biodiversity hotspot, have fragmented forests in their northern parts. This makes these forests additionally vulnerable to Climate Change as well as to increased risk of fire and pest attack. Similarly, forests in parts of western as well as central India are fragmented and have low biodiversity. At the same time, these regions are likely to witness a high increase in temperature and either a decline or marginal increase in rainfall.

Wotkyns, in a report (2011), states:

> Several tribes in Arizona and New Mexico have programs specifically focused on climate change. However, many of those who do not have climate change programs have been doing projects related to climate change mitigation or adaptation within their existing departmental programs, although these projects typically are not identified as "climate change." Other projects by tribes that serve to increase their resiliency to climate change include monitoring of ecosystems, restoration of riparian habitats, removal of invasive species, agricultural initiatives, thinning of forests, and outreach and education. The Pueblo of Zia undertook a project to restore a scared spring that had dried up; the tribe had been concerned that erosion, climate change and the region's growing demand for water would keep the spring from recovering. The Mescalero Apache Tribe has undertaken various fuels treatment projects that move towards thinning forests and utilizing biomass as well as helping support biodiversity within the tribe's forests, and the Santa Clara Pueblo initiated a hazardous fuels reduction project which will reduce its vulnerability to forest fires. The Pueblo of Tesuque has been supporting the Tesuque Farms Agricultural Initiative that has turned 40 acres of land into a productive farm that serves the pueblo and local surrounding communities. The goal of the farm is to help the community become more sustainable, preserve traditional seeds and foods, and maintain a healthier lifestyle, and it is a great source of research and education for the Pueblo.
>
> The Gila River Indian Community, which received a US EPA Climate Showcase Community grant that is funding the tribe's climate change mitigation efforts: Reduction of Greenhouse Gas Emissions

by Development of an Innovative Climate Projects Coordination Structure. The tribe hired a Climate Projects Specialist and is coordinating new and existing project teams, including the Gila River Indian Community Renewable Energy Team that was formally established to implement energy conservation projects. The tribe plans to receive training on LEED certification and promote the program through presentations and articles in local media outlets, complete a community wide greenhouse gas (GHG) inventory, implement a curb-side recycling program, implement a compact fluorescent lighting and Green Building program, and develop options for reducing industrial facilities' GHG emissions.

Some tribes in Arizona and New Mexico have been exploring the opportunities for developing the renewable resources available on their lands. Renewable energy projects include a number of feasibility studies for large-scale wind or solar projects. The Navajo Nation is getting closer to developing its wind energy resources with wind farms, and the Pueblo of Jemez is moving forward with its utility-scale solar project and is also exploring the potential for developing its geothermal energy resources. Other tribes are still in the process of determining the feasibility of renewable energy, have faced challenges to developing the resources, or have determined that the energy resource is inadequate for development. Some tribes are focusing on smaller scale renewable energy projects, such as installing solar panels on tribal buildings as the Hopi Tribe and Fort McDowell Yavapai Nation have done.

Many tribes recently received funding from the US Department of Energy's Energy Efficiency and Conservation Block Grant Program for energy efficiency and weatherization projects. These include projects such as conducting energy audits, installing solar-powered streetlights, upgrading water heaters or installing solar water heaters, retrofitting windows, providing energy-efficient wood stoves, insulating buildings, and developing energy plans.

Tribes in the Southwest and across the nation are taking action on climate change, despite challenges such as limited tribal staff and financial resources. They are trying to find ways to reduce greenhouse gas emissions by implementing renewable energy, energy efficiency, and weatherization measures. They are also trying to maintain the fundamental elements of their cultures in a world that no longer resembles the home of their ancestors. They are undertaking projects that directly or indirectly help their communities and tribal lands adapt and become more resilient to climate change.

Impacts that have already been documented by tribes in the Southwest include erratic weather patterns, including extreme wind events; drought and decreases in water supply; loss of biodiversity and impacts on culturally important native plants and animals;

increases in invasive species; bark beetle damage to forests and increased risk of forest fires; impacts on cattle ranching; and higher utility costs with increased use of air conditioning.

(p. 6)

More area is likely to be afforested under programs such as "Green India mission" and "Compensatory Afforestation Fund Management and Planning Authority" (CAMPA). Therefore, it is necessary to assess the likely impacts of projected Climate Change on existing forests and afforested areas and develop and implement adaptation strategies to enhance the resilience of forests to Climate Change.

The contentious reports of Gadgil and Kasturirangan have percolated so deeply into the existence of these forest communities that most of their precious time is spent traveling to the taluk or district headquarters for Forest Rights Committee meetings. They are yet to get clarity on the matter. In an interview to a prominent freelance environment journalist, Gadgil and Kasturirangan have clearly put the ball in the court of the government. Madhav Gadgil said in an interview:

> Our panel since 2010 has held in-depth consultations with local people. However, the Kasturirangan report claimed that our report was a very rigidly conservation-oriented report and had given no thought to the demands of the local people. This is an absurd claim. Our report must be translated in all state languages and made available to all the Gram Sabhas of the Western Ghats region for their detailed feedback and demarcation of the boundaries of the ecologically sensitive areas.
> (Jamwal, 2014, p. 26)

This statement highlights the tensions between national policies that are based on an "information deficit" model of participation and local research and experience that posits a more complex relationship between individuals and institutions. While this suggests the need to develop more sensitive policies based on the restructuring of socioeconomic and political institutions, this study, based on interactions with key stakeholders, and contexts of media coverage, is skeptical on the role of the power struggle between state and central governments.

Dr. K Kasturirangan said, "after submission of the report to MoEF, the role of the Working Group has come to an end. The limitations are clearly pointed out in the report-as far as the delineation at the village level is concerned" (Jamwal, 2014, p. 31).

The conclusion based on this case study is that greater emphasis must be placed on the negotiation of partnerships that are more sensitive to local diversity and which involve a more equitable distribution of responsibility between different environmental stakeholders. The importance of repeated

Figure 6.1 Snapshot of a Focus Group Discussion (FGD) in progress

exposure to media messages by measuring people's attitudes and beliefs on the issue of Climate Change and then exposing them to new information in the form of television, radio and newspaper reports showing possible future events that illustrated the potential consequences of Climate Change may lead to attitudinal and ultimately behavioral commitment and change, or may inhibit these (Happer and Philo, 2013). People must be able to see real benefits.

Climate Change can be viewed as a factor that is having interdependencies between "actors," who range from governments, to business, to individuals. The contributions of such actors to both the causes and the solutions of Climate Change are vital to boost collective knowledge in understanding past and present climate. This, in turn, becomes key to predicting the future. The vested interests of certain actors in this problem that affects every person on the Earth have a significant impact of the narratives that shape Climate Change discourse (Bushell et al., 2017).

Through these intense deliberations and ethnographic approach, we took the opportunity to create awareness about Climate Change by explaining how the local interpretations and changes that are relatable and narrated by these participants can be connected to science and factual reports in the media. In doing so, the researcher sees this exercise as a humble effort in the direction of establishing vital links between science, policy and media reporting that are crucial for the stakeholders living in these regions to be a part of the solution (see Figure 6.1).

7

REIMAGINING THE NARRATIVE OF CLIMATE CHANGE

"We, the present generation, have the responsibility to act as a trustee of the rich natural wealth for the future generations. The issue is not merely about climate change; it is about climate justice."

~ *Narendra Modi, Prime Minister of India*[1]

International treaties to tackle Climate Change

The SDGs were officially adopted at a UN summit in New York in 2015 and came into effect from January 2016. The deadline for the SDGs is 2030. The SDGs follow and expand on the millennium development goals (MDGs), which were agreed by governments in 2001 and expired in 2015.

The eight MDGs – reduce poverty and hunger; achieve universal education; promote gender equality; reduce child and maternal deaths; combat HIV, malaria and other diseases; ensure environmental sustainability; develop global partnerships – failed to consider the root causes of poverty and overlooked gender inequality as well as the holistic nature of development. The goals made no mention of human rights and did not specifically address economic development. While the MDGs, in theory, applied to all countries, in reality they were considered targets for poor countries to achieve, with finance from wealthy states. Conversely, every country will be expected to work toward achieving the SDGs.

The SDGs are a new, universal set of goals, targets and indicators that UN member states will be expected to use to frame their agendas and political policies over the next 15 years.

Within the 17 goals are 169 targets, which have been developed through a consultative process that brought over 70 national governments and millions of citizens from across the globe together to negotiate and adopt the ambitious agenda.

The UNFCCC is the primary international intergovernmental forum for negotiating the global response to Climate Change.

DOI: 10.4324/9781003015673-7

India and goal 13

India is the fourth-largest emitter of greenhouse gases and is responsible for 5.3% of global emissions. However, the emissions intensity of India's GDP reduced by 12% between 2005 and 2010. In October 2015, India made a commitment to reduce the emissions intensity of its GDP by 20–25% from its 2005 levels by 2020 and by 33–35% by 2030. On 2 October 2016 India formally ratified the historic Paris Agreement. The Government of India has also adopted a National Action Plan on Climate Change to address this issue directly, as well as a National Mission for Green India. These national schemes are complemented by a host of specific programs on solar energy, enhanced energy efficiency, sustainable habitats, water, sustaining the Himalayan ecosystem and encouraging strategic knowledge for Climate Change.

Indian government and its saga with Climate Change

"We are the last generation to address Climate Change" – a profound statement that was amplified and reverberated in every panel discussion at the Clean Cooking Forum 2017, held from 24 to 26 October 2017 in New Delhi.

Home to the world's largest clean cooking gas program, India hosted the biennial Clean Cooking Forum 2017 in October. More than 3 billion people worldwide (800 million in India) are dependent on food cooked over open fires or with heavily polluting solid fuels like wood, charcoal and animal dung. This results in widespread impacts to health, climate, women's empowerment and the environment. In India and China, household air pollution is responsible for up to 30% of outdoor air pollution.

According to an Asian Development Bank report (ADB, 2017), Asia is key to tackle Climate Change. Because of its population, carbon emissions and limited use of renewable energy, Asia's share in global energy–related emissions could reach about 45% in 2030.

"Climate change is a major global challenge. But it is not of our making. It is the result of global warming that came from prosperity and progress of an industrial age powered by fossil fuel," Narendra Modi said while inaugurating the Indian pavilion at the summit, toughening his country's stand in the face of recent US criticism of India. He further stated:

> But we in India face consequences. We see the risk to our farmers. We are concerned about rising oceans that threaten our 7,500 km of coastline and 1,300 islands. We worry about the glaciers that feed our rivers and nurture our civilisation.

Speaking at the UN Climate Change Conference 2015 in Paris, Modi said:

> We assume advanced nations will take ambitious targets. It's not a question of historical responsibility. They also have room for

emission cuts. Climate justice demands with lethal carbon space, developing countries must have enough room to grow.

I have in simple ways stated a dream of new India. I have quoted from 5000-year-old Vedas to say that humans have a right to milk nature but have no right to exploit nature.

~ Prime Minister Narendra Modi, in his address at *St. Petersburg International Economic Forum 2017, Russia*

India's environmental challenges range across many other fronts. Whether it is air, water, waste management, forest cover or resource misuse, the challenge is the same: economic activity has to grow, but the management of its consequences has to improve. This underlines once more that the focus of government activity should be proper regulation and monitoring and the strengthening of the public institutions (like the undermanned state water pollution control boards) that are charged with delivering clean water and air.

The Modi government signaled new ambition when it upped dramatically the target for solar energy capacity. The plan had been to take solar capacity from 3,000 MW in 2012 to 20,000 MW by 2022, which, given the planned sixfold increase, was already considered ambitious. Seeking a change of scale, which is a feature of his approach to his work, Modi has upped the target to 100,000 MW while simultaneously raising the target for wind-generating capacity to 60,000 MW (the country's total generating capacity from all sources is currently 276,000 MW). Actual power generation using the sun and wind will be less than these numbers suggest because of lower capacity utilization than in the case of coal-fired thermal stations. It is not certain that the numbers will be achieved. What is encouraging though is that the cost of solar power has been coming down steadily as the efficiency of solar panels has improved.

The price of solar power can become competitive, especially when compared to the power tariffs paid by industrial and commercial users in most parts of the country. Even if the eventual capacity increase falls short of an outsize target, it would be a significant step forward in adopting low-carbon growth strategies.

If India is to tackle its development and growth objectives without creating an environmental disaster (poisonous air, water scarcity and a breakdown of natural habitats), it has to tailor its development strategy such that nature's assets are protected and harvested sensibly. From the viewpoint of global climate change negotiations too, a low carbon growth strategy is essential. And ambitious clean energy program would obviously make a big difference.

But that is only a small part of what needs to be done. Anyone who pretends that environmental concerns can be pushed into the background while economic growth is made the primary focus is asking for trouble, especially for the poor.

(*Excerpts from* Ninan 2017)

In 2013, the Supreme Court of India, the highest seat of judiciary, observed in an order, "we have realized that forests have the best chance to survive if communities participate in their conservation and regeneration."

The Government of India, in May 2014, changed the nomenclature of its Ministry of Environment and Forests to Ministry of Environment, Forests and Climate Change (MoEFCC) – showing a resolve to understand and adapt to Climate Change. By ratifying the Paris Agreement in 2015, India has shown its global commitment to combat Climate Change while pursuing its development agenda as per its existing laws.

However, long-term solutions are the need of the hour. Adverse weather and uncontrolled pollution are here to stay. But we must take steps to reduce polluting vehicles, burning fossil fuels and burning garbage, to eliminate construction dust and to give farmers incentives so that they do not burn crop residues.

Voicing the stewards of biodiversity – the role of mediators in policy decisions

Most of the households visited on the fringes of major biodiversity reserves do not have access to media – neither newspapers nor television. These also constitute the migratory population, who are dependent on the coffee-picking season and move around for occupation to Kodagu, Chikamagalur and Hassan districts between December and February.

Therefore, our recommendation to the government is that if it wants to conserve these biodiversity hotspots, it must provide access to scientific information. Media should commit to a one-page weekly feature to cover issues pertaining to conservation. This will encourage journalists to report through investigative methods and spur the public to become a part of the solution.

The challenge for developing nations like India is to establish links between development and climate, keeping the SDGs as a framework, to enhance the transition toward a low-carbon economy. A bigger challenge than the aforementioned is to bridge the climate–action gap by implementing forest-related policies. To exemplify, REDD+ could pose to be a constraint for economic and land use factors. Such challenges still predominate even as countries grapple with the challenge of dealing with population and food production (UNEP, 2015).

How can journalists respond to this challenge?

Very often, reporting on Climate Change is apocalyptic, where "doomsday narratives" about what Climate Change will "wreck" serve only to bring about a sense of hopelessness among readers. According to an article published in *The Guardian*, "Feeling hopeless about a situation is cognitively associated with inaction and predicts decreased goal-directed behavior," leading the reader to look the other way. "Climate change adaptation only works when we are hopeful for the future and believe that environmentally vulnerable communities have the agency to act."

This sense of involvement and agency is conveyed in the Citizen Science Project, where scientists track forest pests through people's social media posts. Anderson (2009) says that this type of project makes people feel like they "have a stake," thus encouraging them out of their passivity. He adds that using discourses that leave out explicit references to "Climate Change," while keeping the core message of regulating emissions, has proven effective when addressing people who still have doubts.

A *New York Times* article by Tabuchi tackles this subject by alluding to Trump and how he painted "Democrats as overzealous environmentalists with little sympathy for the economic realities or social mores of rural America." The article states that "climate change discourse is dominated by liberals, which has alienated some conservatives," leading to a trend of "people talking about climate change without talking about it."

For example, the farmer interviewed in the article "can talk for hours about carbon sequestration. . . . Just don't expect him to utter the words climate change." Furthermore, rather than pressing the need to decrease carbon emissions, "regional politicians and business leaders speak of pursuing jobs that clean energy may create," highlighting the importance of adapting rhetoric when addressing different audiences about a politically charged topic.

The challenge of covering Climate Change sometimes runs deeper than words. A BuzzFeed article underlines that debunking viral myths from Climate Change deniers is "near impossible." The article states that a story published by the *Daily Mail* claiming that "world leaders were duped into spending billions over manipulated global warming data" was shared thousands of times. Material questioning the bogus story was shared only one quarter as much. "Because of our cognitive biases, once your opinions are formed it's hard to change that," concludes the BuzzFeed article.

Climate feedback is an initiative to keep an eye on when it comes to verifying information published about Climate Change.

When it comes to working with environmental data, journalists and scientists alike seem to be facing challenges. The main issue seems not to come from scarcity of data but rather to be the lack of journalistic tools available to process this data. Data journalism is the buzzword today, and journalists can use tools (www.globalforestwatch.org/; http://climatetracker.org/) having interactive maps, technology and media convergence techniques (apps that show rise in temperature, rainfall), and data from local scientists and local weather monitoring stations.

Scientists and journalists should therefore be encouraged to work hand in glove more often, and environmental data should be more accessible and easier to interpret. However, the existing incentive structure makes that hard.

According to Anderson (2009), one of the questions data journalists need to ask themselves is: "What does the data say about your city and your life?" Researchers from the University of Austin are precisely trying to this – using the data from their city to give insights to the local population about the environment in their surroundings – through a study on air pollution in Oakland. Google Street View cars drove 15,000 miles, gathering three million unique data sets on air pollution, which were visually displayed to draw people's attention to pollution hotspots within the city. "People care about the air they breathe, and [pollution] is a massive health crisis," underlines Anderson.

Anderson underlines that a significant challenge in reporting on Climate Change is the fact that in the early days, reporting tended to emphasize that the most serious effects will be centuries away, which in turn put readers into a false sense of security.

The Guardian, for example, uses data in a more dynamic way, as evidenced through its use of an interactive tool dating back to 2015 that shows how much fossil fuel has been extracted in the world since the reader has arrived on the page. The reader is also given the option of entering their age to show how old they will be when the global carbon budget is blown. The interaction therefore serves as a quasi-call to action, illustrating the speed at which changes are occurring, thus making Climate Change a lot more relatable.

Climate Change is not an easy topic to report on. Some of the challenges include making scientific jargon relatable to the public and simplifying a complex issue that is wrapped up in politics with a plethora of long-term and short-term effects on society – not to mention Climate Change skeptics and the viral spreading of myths, which only serve to complicate reporting further. Foxwell-Norton and Lester who studied the "Save the Reef" campaign said: "media and communications should present unprecedented power for civil society actors to impact upon formal processes of definitions and decision-making" (2017, p. 579).

Last but not the least, journalists and media houses must accept that Climate Change is a challenging and complex problem, which they have to navigate through to get the real picture out – lucidly and with the help of data. The media have to rise above the complexities of the issue and urge the government to support policies even if they come at a personal cost to them. The pressure can be built by informing the public, including big business houses, small farmers, celebrities, teachers, not-for-profit organizations and climate scientists.

By using the Western Ghats as a case study, we examined media representations of Climate Change through media coverage of the declaration of the Western Ghats as one of the world's biodiversity hotspots by UNESCO. The findings provide evidence that coverage of issues related to Climate Change peak during times of crisis. Media coverage has also been linked to the controversy surrounding the zoning of ESAs in the Karnataka region of the Western Ghats. It can be concluded that media coverage peaks when issues related to overlapping themes of sustainability, policy decisions and Climate Change occur.

Next, we highlight key findings of our study to further throw light on how this particular case study of analysis of news coverage with reference to declaration of Western Ghats as an Ecologically Sensitive Area can be viewed in the larger context on coverage related to Climate Change.

A glance of the media coverage

- There were 552 articles across the five-year study period in the two publications that made significant reference to Climate Change in the Western Ghats.
- Forty-one percent of articles (226) made significant mention of Western Ghats in the context of Climate Change policy.
- Fifty-nine percent of articles (326) across the two publications that made significant mention of Climate Change were news reports.
- *TOI* had the highest proportion (83.7%) of articles that were 500 words or longer and contained analytical essays (69.6%), thereby providing better perspectives and wider sources about Climate Change.

Genre of Climate Change articles

- The two newspapers were analyzed for their coverage of reports related to Climate Change and the Western Ghats. The genres created for classification are reports related to the declaration of the Western Ghats as an ESA, articles on conservation, articles on sustainability, articles on eco-tourism, articles on forest communities and articles on resources/energy in the Western Ghats (hydel projects, mining).
- *Deccan Herald gave* more prominence (72%) to ESA reports in its coverage, and
  TOI (97%) focused largely on sustainable issues during June 2012–2017.
- *TOI* (65%) showed a concern for all angles related to the Western Ghats (threat to the ecosystem, threat to the communities and threat to the economy) in its coverage.
- *Deccan Herald*, especially in 2013, after the declaration of the Western Ghats as a biodiversity site by UNESCO, published more articles (64%) related to conservation of the fragile ecosystem.

Placement of Climate Change coverage

The prominence given to coverage of the most pressing issue of this century did not reflect in the editorial decisions of the two newspapers.

- A major portion of articles (96%) were placed in the inside pages of the newspapers.
- There were nine articles that made it to the front page and ten articles that were editorials in both the publications.
- *TOI* was more likely to publish front-page articles (77.8%).

Main sources

- The two leading publications in the state of Karnataka published articles (42%) that had an official version as the main source.
- The two newspapers turned to environmental experts as their second main source for articles related to the Western Ghats and Climate Change (25%).
- There were 52.5% articles by *TOI* that made significant reference to Climate Change in the Western Ghats through environmental experts as the main source.
- There were 47.5% articles by *Deccan Herald* that made significant reference to Climate Change in the Western Ghats through environmental experts as the main source.

Representations of the climate science–policy nexus from 2012 in the Indian newspapers analyzed in this study are evidence of a new sense of urgency attached to the risks from Climate Change. News reports show a tendency on the part of the journalists to talk about issues like the interests and commitments of actors involved in the intersection between conservation and development, contributing to a more in-depth understanding of the politics of Climate Change (Carvalho and Burgess, 2005).

The bitter battle in the states of Kerala, Karnataka, Goa and Maharashtra over the Gadgil and Kasturirangan reports on the conservation of the Western Ghats has been receiving widespread media coverage. The exclusive focus on "Ecologically Sensitive Areas" and the protests and efforts to exclude certain areas as ESAs have taken the focus off the larger debate on sustainable development and conservation (Jamwal, 2014; Nair and Moolakkattu, 2017).

The Karnataka Government opposed this recognition, claiming that the forest dwellers will suffer as a result of this act, and that it would hinder 'development programs'. Subsequently the Government of Karnataka (Forests come under the State's purview) released a report (Kasturirangan report) – the Kasturirangan committee was set up to "amend" the findings of the Gadgil committee – which was seen as too "anti-development" – Madhav

Gadgil went on to the extent of calling Kasturirangan's report "undemocratic". "the key issue is to understand the ecologically sensitive area in the Western Ghats which raised issues about illegal mining and other activities in many regions along the Western Ghats," explains Lingaraj Jayaprakash, doctoral candidate, McGill University, Canada, working on environmental policy issues, who was an expert member for this study, while reiterating that "an ecological disturbance in the Western Ghats will have a developmental impact across much of south India."

As validated by the expert Lingaraj Jayaprakash, the media reports in this study invariably pointed toward the contestations to the recommendations of Kasturirangan and Gadgil reports from various quarters without any explicit Climate Change connection but with a relevance to the fate of policy, thereby chronicling the futility of the policy process (Jogesh, 2011). Another important source suggested by the expert panelist Lingaraj Jayaprakash (see Appendix B) was MLA/MP questions (during the zero hour) raised in the Legislative Assembly/Parliament on Western Ghats–related issues.

Journalists are now addressing causal links between extreme "natural events" and are leaning toward an interpretative tendency reinforced by data and expert referencing in their reports. Part of their challenge is that the world's media need – and use – overarching narratives to describe the Climate Change "mega-story." Alarming stories of famine, sea level rises, floods, hurricanes and droughts easily grab attention. While mainstream media are likely to steer clear of reporting that deeply challenges corporate interests that have a strong connection with their outlets, it would be far too simplistic to suggest that ties with fossil fuel industries completely prevent critical reporting (Anderson, 2009).

Boykoff (2011) expresses sympathy for journalists, who need "to focus and contextualize a story with tight deadlines." He appeals for a more educational role for journalism but is realistic enough to see that increased "climate literacy" is not going to curb the trend on its own (Boykoff, 2011). As Boykoff and Boykoff (2007) observe, the trajectory of global warming coverage does not follow a "natural" cycle and is strongly connected to policy developments.

This study, in order to validate the research objectives, did not try to seek correlation, as this would not really inform much about policy and practice. As pointed out by one of the expert panelists, Lingaraj, the study, however, did find some correlation. However, the main objective was to distinguish the exact nature and mechanisms through which media impacted policy and conservation efforts. In doing so, the study identified the key social actors (official sources and environmental experts) and media outlets, as well as the interests driving this spike in coverage.

The study was also able to point toward the frames that emerged from these media reports and found that there existed a mismatch among media reporting, scientific opinion and local stakeholder concerns.

Media reporting on Climate Change is still in its infancy, especially in the context of the Western Ghats. The impact of media reporting on conservation and policy is limited in the media but definitely exists. The late Rohith B.R., Assistant Editor, *TOI*, who had been reporting on issues conserving the Western Ghats and the environment in Karnataka state for the *TOI Karnataka Edition*, said:

> The Government of Karnataka is yet to submit its stand on declaration of ESAs, as there are different formulas, yardsticks with regard to the ESAs in all the states. Meanwhile, the carrying capacity of river Cauvery has reduced due to a decrease in forest cover.[2]

This "disaster" angle is by far the most common one in the coverage of Climate Change, as shown by several studies. At times, this "alarming" story morphs into the more "alarmist" language of catastrophe, calamity or doom. A study carried out by the Reuters Institute for the Study of Journalism shows that in the television coverage of the three recent blockbuster reports by the IPCC, the disaster narrative was still by far the most common in the six countries it examined. The study examined the coverage on television, which is still in most countries the most used and trusted source of information for news in general and for news about science (Painter, 2013).

As *Columbia Journalism Review* noted, a Climate Change report helped to shift the nature of the Climate Change story in the media. It became a business story on the business pages, reaching a new and powerful audience (Eshelman, 2014). The Western Ghats, in addition to being a biodiversity hotspot, is home to several commercial crops such as coffee, tea, pepper, cardamom, vanilla and honey, contributing significantly to the economy. Hydel power projects and small and large infrastructure projects, including laying railway lines and high-tension power lines, are a major source of revenue for the government. Tourism is another thriving industry in this region. The policy implications after the declaration by UNESCO is seen as a major impediment for development in this region, and the media is playing its part here in partly fueling the debate, partly acting as a moderator and sometimes "sitting on the fence" between the policy makers and public.

Research shows that the promotion or presence of uncertainty in media reporting of climate science can act as an obstacle to public understanding and lead to disengagement, so it is critically important that newspapers provide proper context when covering such an important issue (Happer and Philo, 2013).

In the light of its potential benefits, some scientists have been using the concept of risk to frame their discussions of Climate Change. At the moment, the media hardly pick up on risk language (Painter, 2015). This case study and an analysis of five years of discourse on Climate Change demonstrates that media representation is an important agenda-setting factor for audiences and is a significant influence in shaping people's knowledge and perceptions of the issue. Coverage of Climate Change is strongly linked to policy issues and political agenda on this issue (Carvalho and Burgess, 2005).

We believe this study has contributed to both climate communication and journalism studies and has been unique to combine three crucial actors in this performance – public, policy makers and media. While doing so, this study also touches upon the cultural contexts that lead media houses to frame the argument and context to suit their editorial stance (Anderson, 2009).

Deccan Herald is seen as a more local and regional representative for the people, and its coverage was focused on the policy decisions and their implications on the Western Ghats. *TOI*, a national daily, focused its coverage on events related to Climate Change environmental summits at the international level, rather on regional and local issues.

Climate Change is the most pressing issue of the twenty-first century. Human-induced Climate Change, according to scientists, could push more than 100 million people into extreme poverty by 2030. Asia holds the key to tackle Climate Change because of its population, rapid rate of carbon emissions and ability to use renewable energy, due to its tropical climate.

However, the bigger challenge is to influence policy decisions and the public by shaping information dissemination about Climate Change. Breaking down complex scientific estimates about Climate Change into solvable components that can be synthesized by the public is a big challenge for the media.

This study shows that first, the reporting on Climate Change by the media is largely driven by two forces – a pathbreaking scientific breakthrough and a major policy decision impacting the lives of its readership. Through an analysis of Indian newspaper coverage during 2012–2017, this study finds that the Indian media is yet to understand the significance of its role as a facilitator between scientists, policy makers and the public. The results clearly show that official sources dominate in the media's narrative of environmental coverage. Second, official sources are a large and heterogeneous category, suggesting that it is time to move beyond a focus on the dominance of official sources per se and toward research and theory that explores differences within the category of official sources and their antecedents. Third, journalists face multiple challenges from official sources in communicating Climate Change. Finally, convincing the public to take steps to mitigate Climate Change was the most widely used rhetorical device by journalists in this study.

Given that Climate Change is here to stay, this study gains significance as it emphasizes that the media assume a greater role in connecting climate science to policy decisions and the public. Finally, the results point toward journalistic biases based on their own organizational and national influences on media content. In maintaining the balance between conservation efforts and development, climate journalism must interact and complement these different factors to ultimately empower the public to become a part of the solution and the policy-making process.

On 1 July 2012, after a six-year-long campaign, 39 sites in the Western Ghats made it to the World Heritage Sites list of UNESCO, demonstrating the need and commitment for conservation of this rich biodiversity hotspot. As noted by the committee, these sites are "ongoing ecological and biological processes in the evolution and development of terrestrial, freshwater, coastal and marine ecosystems and communities of plants and animals" and are crucial for the survival of our planet. In this book, the Western Ghats are a macrocosm to understand Climate Change and the role played by Press narratives. Through this macrocosm it is possible to look at the biodiversity crisis that is staring at humanity threatening to throw economies and societies into crisis. The recent Intergovernmental Panel on Climate Change – an authoritative body for collating scientific evidence on the state of climate emergency – has published its report and it makes for very sober reading. We already know that we are in the middle of catalytic shifts occurring deep within structural changes of our climate. Indigenous people, their local wisdom, can complement scientific understanding. We have seen evidence of this historically, especially in places where instrumental observations were sparse (IPCC, 2021). The media must effectively engage and have further dialogue with these marginalized and indigenous communities (constituting 5% of the world's population) who can become credible sources, to create an indelible impact on public and political discourse. Like the image used to depict Climate Change, on the cover page of *Time* Magazine in September 2000 (Polar Bear on an isolated shelf of ice), that warned "Arctic Meltdown" two decades ago, the media's resurgent moment is now. The last shelf of melting ice that media industry can hold onto and hopefully turn the tide, thus becoming a key pillar to lead us toward Collective Climate Action, instead of contributing to a global catastrophe, is to increase all resources it has toward coverage of solutions to mitigate the effects of Climate Change.

Notes

1 Quoted in Ganguly, A., and Chauthaiwale, V. (2016). *The Modi doctrine: New paradigms in India's Foreign policy* (p. 5). India: Wisdom Tree Publishers, New Delhi.
2 Personal interview with Rohith B R in 2017.

REFERENCES

ADB. (2017). *'Poverty in India.' Basic statistics*. Asian Development Bank. Retrieved October 19, 2017, from www.adb.org/countries/india/ poverty

Allan, J. R., Venter, O., Maxwell, S., Bertzky, B., Jones, K., Shi, Y., and Watson, J. E. (2017). Recent increases in human pressure and forest loss threaten many natural world heritage sites. *Biological Conservation, 206*, 47–55. doi:10.1016/j. biocon.2016.12.011

Altheide, D. L., and Schneider, C. J. (2013). *Qualitative media analysis* (2nd ed.). London: Sage.

Analysis: The climate papers most featured in the media in 2015. (2016, January 18). Retrieved from www.carbonbrief.org/analysis-the-climate-papers-most-featured-in-the-media-in-2015

Anderegg, W., et al. (2010). Expert credibility in climate change. *Proceedings of the National Academy of Sciences*. doi:10.1073/pnas.1003187107. Retrieved from http://tinyurl.com/AndereggPNAS

Anderson, A. (2009). Media, politics and climate change: Towards a new research agenda. *Sociology Compass, 3*(2), 166–182. doi:10.1111/j.1751-9020.2008.00188.x

Antilla, L. (2005). Climate of scepticism: US newspaper coverage of the science of climate change. *Global Environmental Change: Human and Policy Dimensions, 15*(4), 338–352.

Audit Bureau of Circulation. (2017). Retrieved September 1, 2017, from www. auditbureau.org

Bapuji Rao, B., Santhibhushan Chowdary, P., Sandeep, V. M., Rao, V. U. M., and Venkateswarlu, B. (2014). Rising minimum temperature trends over India in recent decades: Implications for agricultural production. *Global Planet Change, 117*, 1–8.

Becker, J., and Lissman, H. J. (1973). *Inhaltsanalyse – Kritik einer sozialwissenschaftlichen Methode*. Arbeitspapiere zur politischen Soziologie 5. München: Olzog.

Bhagwat, S., Kushalappa, C., Williams, P., and Brown, N. (2005). A landscape approach to biodiversity conservation of sacred groves in the Western Ghats of India. *Conservation Biology, 19*, 1853–1862.

Billet, S. (2010). Dividing climate change: Global warming in the Indian mass media. *Climatic Change, 99*(1/2), 525–537.

Bjola, C., and Holmes, M. (2015). *Digital diplomacy: Theory and practice*. London, UK: Routledge.

Blumler, J. G., and Katz, E. (1974). The uses of mass communications: Current perspectives on gratifications research. *Sage Annual Reviews of Communication Research*, 3. https://eric.ed.gov/?id=ED119208

Bose, P. (2017). Climate adaptation: Marginal populations in the vulnerable regions. *Climate and Development*, 9(6), 575–578.

Boykoff, M., Andrews, K., Daly, M., Katzung, J., Luedecke, G., Maldonado, C., and Nacu-Schmidt, A. (2018). *A Review of Media Coverage of Climate Change and Global Warming in 2017, Media and Climate Change Observatory, Center for Science and Technology Policy Research, Cooperative Institute for Research in Environmental Sciences*, University of Colorado. Retrieved from http://sciencepolicy.colorado.edu/icecaps/research/media_coverage/summaries/special_issue_2017.html

Boykoff, M. T. (2011). *Who speaks for the climate?* New York, NY: Cambridge University Press.

Boykoff, M. T. (2007). Flogging a dead norm? Media coverage of anthropogenic climate change in United States and United Kingdom from 2003 to 2006. *Area*, 39, 470–481.

Boykoff, M. T., and Boykoff, J. M. (2004). Balance as bias: Global warming and the US prestige press. *Global Environmental Change*, 14, 125–136.

Boykoff, M. T., and Boykoff, J. M. (2007). Climate change and journalistic norms: A case study of US mass-media coverage. *Geoforum*, 38, 1190–1204.

Boykoff, M. T., and Roberts, J. T. (2007). *Media coverage of climate change: Current trends, strengths, weaknesses.* Human Development Report 2007/8 – United Nations Development Programme Occasional paper. New York, NY: Human Development Report Office.

Brainard, C. (2008, November 26). Global cooling, confused coverage. *Columbia Journalism Review*.

Briggs, C. L., and Hallin, D. C. (2007). Biocommunicability: The neoliberal subject and its contradictions in news coverage of health issues. *Social Text*, 25(4), 43–66.

Briggs, C. L., and Hallin, D. C. (2016). *Making health public: How news coverage is remaking media, medicine, and contemporary life.* London and New York, NY: Routledge.

Brüggemann, M., and Engesser, S. (2017). Beyond false balance: How interpretive journalism shapes media coverage of climate change. *Global Environmental Change*, 42, 58–67. doi:10.1016/j.gloenvcha.2016.11.004

Brüggemann, M., Silva-Schmidt, F., Hoppe, I., Arlt, D., and Schmitt, J. B. (2017). The appeasement effect of a United Nations climate summit on the German public. *Nature Climate Change*, 7, 783–787. doi:10.1038/nclimate3409.

Burgess, J. (1990). The production and consumption of environmental meanings in the mass media: A research agenda for the 1990s. *Transactions of the Institute of British Geographers*, 15(2), 139–161.

Bushell, S., Buisson, G. S., Workman, M., and Colley, T. (2017). Strategic narratives in climate change: Towards a unifying narrative to address the action gap on climate change. *Energy Research & Social Science*, 28, 39–49.

Carvalho, A. (2007). Ideological cultures and media discourses on scientific knowledge: Re-reading news on climate change. *Public Understanding of Science*, 16, 223–243.

Carvalho, A., and Burgess, J. (2005). Cultural circuits of climate change in U.K Broadsheet newspapers, 1985–2003. *Risk Analysis, 25*(6), 1454–1469.

Chapman, G., Kumar, K., Fraser, C., and Gaber, I. (1997). *Environmentalism and the mass media: The Northsouth divide* (1st ed.). New York, NY: Routledge.

Chaturvedi, R. K., Gopalakrishna, R., Jayaraman, M., Bala, G., Joshi, N. V., Sukumar, R., and Ravindranath, N. H. (2011). Impact of climate change on Indian forests: A dynamic vegetation modeling approach. *Mitigation and Adaptation Strategies for Global Change, 16*, 119–142.

Choudhury, C. (2016, April 07). Making a hollow in FRA. *The Hindu.*

Climate Change threatens parasites. (2017, September 26). *Deccan Herald*, p. 4.

Coffee Board of India. (2017). *Coffee statistics*. Retrieved from www.indiacoffee. org/coffee-statistics.html

Coleman, S., and Blumler, J. G. (2009). *The Internet and democratic citizenship: Theory, practice and policy.* New York, NY: Cambridge University Press.

Cook, J., Nuccitelli, D., Green, S. A., Richardson, M., Winkler, B., Painting, R., . . . Skuce, A. (2013). Quantifying the consensus on anthropogenic global warming in the scientific literature. *Environmental Research Letters, 8*(2), 24024. doi:10.1088/1748-9326/8/2/024024

Craparoa, A. C. W., Van Astenb, P. J. A., Läderachc, P., Jassogneb, L. T. P., and Graba, S. W. (2015). Coffee arabica yields decline in Tanzania due to climate change: Global implications. *Agricultural and Forest Meteorology, 207*, 1–10.

D'Angelo, P. (2002). News framing as a multiparadigmatic research program: A response to Entman. *Journal of Communication, 52*, 870–888.

Deforestation contributes more. (2017, September 19). *Deccan Herald*, p. 4.

Entman, R.M. (1993). Framing: Toward clarification of a fractured paradigm. *Journal of Communication, 43*(4), 51–58.

Entman, R. M., and Rojecki, A. (1993). Freezing out the public: Elite and media framing of the U.S. anti-nuclear movement. *Political Communication, 10*(2), 151–167.

Eshelman, R. S. (2014). Risky business. *Columbia Journalism Review*. Retrieved November 15, 2017, from http://archives.cjr.org/the_observatory/climate_change_ risky_business.php

Foxwell-Norton, K., and Lester, L. (2017). Saving the great barrier reef from disaster, media then and now. *Media, Culture & Society, 39*(4), 568–581.

Ganapathy, D. (2015). *Content Analysis of tweets by Indian Politicians and Celebrities* (PhD thesis). University of Mysore, India.

Ganapathy, D. (2017, January 10). Bees feel the sting of climate change. *Deccan Herald*, p. 4. Retrieved from www.deccanherald.com/content/590646/bees-feel-sting-climate-change.html.

Hallin, D. C., Brandt, M., and Briggs, C. L. (2013). Biomedicalization and the public sphere: Newspaper coverage of health and medicine, 1960s – 2000s. *Social Science and Medicine, 96*, 121–128.

Han, G., Chock, T. M., and Shoemaker, P. J. (2009). Issue familiarity and framing effects of online campaign coverage: Event perception, issue attitudes, and the 2004 presidential election in Taiwan. *Journalism & Mass Communication Quarterly, 86*(4), 739–755.

Han, G., and Wang, X. (2012). Understanding 'made in China' valence framing and product- country image. *Journalism & Mass Communication Quarterly, 89*(2), 225–243.

Happer, C., and Philo, G. (2013). The role of the media in the construction of public belief and social change. *Journal of Social and Political Psychology, 1*(1), 321–336, doi:10.5964/jspp.v1i1.96

Howland, D., Becker, M. L., and Prelli, L. J. (2006). Merging content analysis and the policy sciences: A system to discern policy-specific trends from news media reports. *Policy Science, 39,* 205–231.

IEA. (2016). *World energy outlook 2016.* Paris, France: International Energy Agency.

IPCC. (2007). Climate change: The physical science basis. In S. Solomon, D. Qin, M. Manning, Z. Chen, M. Marquis, K. B. Averyt, M. Tignor, and H. L. Miller (Eds.), *Contribution of working group I to the fourth assessment report of the intergovernmental panel on climate change.* Cambridge and New York, NY: Cambridge University Press.

IPCC. (2014a). Climate change 2014: Synthesis report. Contribution of working groups I, II and III to the fifth assessment report of the intergovernmental panel on climate change. In R. K. Pachauri and L. A. Meyer (Eds.), *Core writing team.* Geneva, Switzerland: IPCC.

IPCC. (2014b). Summary for policymakers. In O. Edenhofer, R. Pichs-Madruga, Y. Sokona, E. Farahani, S. Kadner, K. Seyboth, A. Adler, I. Baum, S. Brunner, P. Eickemeier, B. Kriemann, J. Savolainen, S. Schlömer, C. V. Stechow, T. Zwickel, and J. C. Minx (Eds.), *Climate change 2014: Mitigation of climate change.* Contribution of Working Group III to the Fifth Assessment Report of the Intergovernmental Panel on Climate Change. Cambridge and New York, NY: Cambridge University Press.

IPCC. (2021). *Climate Change 2021: The physical science basis.* Contribution of Working Group I to the Sixth Assessment Report of the Intergovernmental Panel on Climate Change [Masson-Delmotte, V., P. Zhai, A. Pirani, S.L. Connors, C. Péan, S. Berger, N. Caud, Y. Chen, L. Goldfarb, M.I. Gomis, M. Huang, K. Leitzell, E. Lonnoy, J.B.R. Matthews, T.K. Maycock, T. Waterfield, O. Yelekçi, R. Yu, and B. Zhou (eds.)]. Cambridge University Press. In Press.

Jamwal, N. (2014, February). Report wars. *Eco Guru,* pp. 26–31.

Jayakumar, M., Rajavel, M., and Surendran, U. (2017). Impact of climate variability on coffee yield in India-with a micro-level case study using long-term coffee yield data of humid tropical Kerala. *Climatic Change.* https://doi.org/10.1007/s10584-017-2101-2

Jogesh, A. (2011). A change in climate? Trends in climate change reportage in the Indian print media. In N. K. Dubash (Ed.), *Handbook of climate change and India.* doi:10.4324/9780203153284.ch20

Krippendorff, K. H. (2012). *Content analysis: An introduction to its methodology* (3rd ed.). Thousand Oaks: SAGE Publications, Inc.

Kuypers, J. A. (2002). *Press bias and politics: How the media frame controversial issues* Westport, CT: Praeger Publishers.

Lele, S. (2017). Forest governance from co-option and conflict to multilayered governance? *Economic & Political Weekly, LII*(25 & 26), 55–58.

Lombard, M., Snyder-Duch, J., and Brack, C. (2002). Content analysis in mass communication: Assessment and reporting of intercoder reliability. *Human Communication Research, 28*(4), 587–604.

Macchi, M., Gurung, M. A., and Hoermann, B. (2017). Community perceptions and responses to climate variability and change in the Himalayas. *Climate and Development, 7*(5), 414–425.

Mayring, P. (2000). Qualitative content analysis. *Forum: Qualitative Social Research*, 1(2). Retrieved from http://217.160.35.246/fqs-texte/2-00/2-00mayring-e.pdf

McComas, K., and Shanahan, J. (1999). Telling stories about global climate change measuring the impact of narratives on issue cycles. *Communication Research*, 26(1), 30–57.

McCombs, M. E. (2005). A look at agenda-setting: Past, present, and future. *Journalism Studies*, 6, 543–557.

McQuail, D. (2005). *McQuail's mass communication theory*. London: Sage.

MoEF. (2009). *Diversion of forestland for non-forest purposes under the forest (conservation) act, 1980*. Ensuring Compliance of the Scheduled Tribes and Other Traditional Forest Dwellers (Recognition of Forest Rights) Act 2006, Circular No: F No 11–9/1998 – FC (pt), Ministry of Environment and Forests (FC Division), Government of India, New Delhi.

MoEF. (2013). *Report of the high level working group on Western ghats*. High Level Working Group Report on Western Ghats – reg. | The Official Website of Ministry of Environment, Forest and Climate Change, Government of India. Retrieved from moef.gov.in.

Nacu-Schmidt, A., Oonk, D., Pearman, O., Boykoff, M., Daly, M., McAllister, L., and McNatt, M. (2017). *Indian newspaper coverage of climate change or global warming, 2000–2017*. Center for Science and Technology Policy Research, Cooperative Institute for Research in Environmental Sciences, University of Colorado, Web. Retrieved November 14, 2017, from http://sciencepolicy.colorado.edu/media_coverage

Nagarajan, S., Jagadish, S. V. K., Hari Prasad, A. S., Thomar, A. K., Anand, A., Pal, M., and Agarwal, P. K. (2010). Local climate affects growth, yield and grain quality of aromatic and non-aromatic rice in northwestern India Agric. *Ecosystems & Environment*, 138, 274–281.

Nair, N. V., and Moolakkattu, J. S. (2017). The Western Ghats imbroglio in Kerala a political economy perspective. *Economic & Political Weekly*, LII(34), 56–65.

Nayak, N. (2017). Climate change policy, federalism and developmental process in India. *International Journal of Advance Research and Innovative Ideas in Education*, 3(2), 37–47.

Neuendorf, K. A. (2002). *The content analysis: Guide book*. Thousand Oaks, CA: SAGE.

Newman, T. P. (2016). Tracking the release of IPCC AR5 on Twitter: Users, comments, and sources following the release of the Working Group I Summary for Policymakers. *Public Understanding of Science*, 1–11. doi: 10.1177/0963662516628477.

Ninan, T. N. (2017). *Turn of the tortoise: The challenge and promise of India's future*. New York, NY: Oxford University Press. Retrieved from www.downtoearth.org.in/reviews/growth-or-environment-52481

O'Neill, S., and Nicholson-Cole, S. (2009). "Fear won't do it" promoting positive engagement with climate change through visual and iconic representations. *Science Communication*, 30(3), 355–379. https://doi.org/10.1177/1075547008329201

Oreskes, N. (2004). Beyond the ivory tower: The scientific consensus on climate change. *Science*, 306(5702), 1686–1686. doi:10.1126/science.1103618

Painter, J. (2007). *All doom and gloom?* International TV Coverage of the April and May 2007 IPCC Reports.

Painter, J. (2010). *Summoned by science: Reporting climate change at Copenhagen and beyond*. Oxford: Reuters Institute for the Study of Journalism, University of Oxford.

Painter, J. (2011). *Poles apart the international reporting of climate scepticism.* Oxford: Reuters Institute for the Study of Journalism, University of Oxford.

Painter, J. (2013). Climate change in the media: Reporting risk and uncertainty. Oxford: Reuters Institute for the Study of Journalism, University of Oxford.

Painter, J. (2015, March 25). 'Taking a bet on risk'. Commentary. *Nature Climate Change,* 5, 286–288. https://doi.org/10.1038/nclimate2542

Painter, J., Kristiansen, S., and Schäfer. M. S. (2018). How 'Digital-born' media cover climate change in comparison to legacy media: A case study of the COP 21 summit in Paris. *Global Environmental Change,* 48, 1–10.

Pittock, A. B. (2009). *Climate change: The science, impacts and solutions* (2nd ed.). Collingwood: CSIRO Publishing.

Poornananda, D. S. (2017). Development – induced displacement and the print media in Karnataka. *Journal of Media and Social Development,* 5(1), 5–44.

Rohith, B. R. (2017, January 24). Research shows butterflies can act as bio-indicators of climate change. *The Times of India,* p. 2.

Rosas-Moreno, T., and Ganapathy, D. (2021). Has India's tripartite cooperation with Brazil and South Africa helped it combat human trafficking? A news media framing analysis spanning two decades. *Journalism,* 22 (7), 1831–1850. First Published February 26, 2019. https://doi.org/10.1177/1464884919831093

Sampei, Y., and Aoyagi-Usui, M. (2009). Mass-media coverage, its influence on public awareness of climate-change issues, and implications for Japan's national campaign to reduce greenhouse gas emissions. *Global Environmental Change,* 19, 203–212.

Schreier, M. (2012). *Qualitative content analysis in practice.* Thousand Oaks, CA: Sage.

Smith, P., Bustamante, M., Ahammad, H., Clark, H., Dong, H., Elsiddig, E., . . . Tubiello, F. (2014). Agriculture, forestry and other land use (AFOLU) In *Climate change 2014: Mitigation of climate change.* Contribution of Working Group III to the Fifth Assessment Report of the Intergovernmental Panel on Climate Change. Cambridge and New York: Cambridge University Press.

Snow, D. A., and Benford, R. D. (1988). Ideology, frame resonance, and participant mobilization. *International Social Movement Research,* 1, 197–217.

Social Media Analytics for Digital Advocacy Campaigns: Five Common Challenges (Publication). (2016). *USC centre on public diplomacy.*

UNEP. (2014). *The emissions gap report 2014.* Nairobi: United Nations Environment Programme (UNEP). Retrieved October 24, 2017, from http://wedocs. unep.org/bitstream/handle/20.500.11822/9345/-The%20Emissions%20Gap%20 Report%202014%3a%20a%20UNEP%20 synthesis%20report-November%20 2014EGR_2014_Lowres.pdf?sequence=3&isAllowed=y

UNEP. (2015). *The emissions gap report 2015.* Nairobi: United Nations Environment Programme (UNEP). Retrieved October 24, 2017, from https://uneplive. unep.org/media/docs/theme/13/EGR_2015_301115_lores.pdf

UNEP. (2017). *The emissions gap report 2017.* Nairobi: United Nations Environment Programme (UNEP). Retrieved October 24, 2017, from www.unenvironment.org/resources/emissions-gap-report

UNESCO. (2013). *Climate change in Africa: A guidebook for journalists.* Retrieved November 2, 2017, from http://unesdoc.unesco.org/images/0022/002254/ 225451e.pdf

UNFCCC. (2017). Retrieved November 2, 2017, from http://unfccc.int/paris_agreement/ items/9485.php

Vachani, S., and Usmani, J. (Eds.). (2014). *Adaptation to climate change in Asia.* Cheltenham, UK: Edward Elgar.

Van Gorp, B. (2005). Where is the frame? Victims and intruders in the Belgian press coverage of the asylum issue. *European Journal of Communication, 20*(4), 484–507.

Weart, S. (2012, August 17). *The discovery of global warming [excerpt].* Retrieved November 15, 2017, from www.scientificamerican.com/article/discovery-of-global-warming/

Wilkins, L. (1993). Between facts and values: Print media coverage of the greenhouse effect, 1987–1990. *Public Understanding of Science, 2,* 71–84.

Wotkyns, S. (2011). *Tribal climate change efforts in Arizona and New Mexico.* Flagstaff, AZ: Institute for Tribal Environmental Professionals.

Young, N., and Dugas, E. (2012). Comparing climate change coverage in Canadian English-and French-language print media: Environmental values, media cultures, and the narration of global warming. *Canadian Journal of Sociology, 37*(1), 24–55.

Zaharna, R. S., Arsenault, A., and Fisher, A. (2014). *Relational, networked and collaborative approaches to public diplomacy: The connective mindshift.* New York, NY: Routledge.

Zehr, S. C. (2000). Public representations of scientific uncertainty about global climate change. *Public Understanding of Science, 9,* 85–103.

APPENDIX

Appendix A: codebook content analysis

The coding was performed through an Excel spreadsheet. The key reasons for this decision were to reduce the number of errors made and for speeding up the coding process. The code book was designed in the following manner:

Column A: article number
Column B: date on which published
Column C: word count of article
Column D: place mentioned (to show edition and location)
Column E: section where it appeared in the newspaper (international, national, regional)
Column F: byline
Column G: headline
Column H: article type

Here, we indicate the type of the news article. Categories include:

1   News report
2   Investigative piece
3   Feature
4   Travel/adventure
5   Editorial
6   Infographic
7   Analytical essay

Column I: placement

Here, we indicate the placement of the news article in the newspaper. Categories include:

1 Front page
2 Inside page
3 Editorial page

Column J: main source

Here, we identify the person, group or individual cited the most. Categories include:

1   Law enforcement or official document[1]
2   Legal agency or representative/document
3   Nongovernmental agency or NGO representative/document
4   UN
5   Citizen
6   Environmental expert

Column K: tone of the article

Here, we indicate if the general gist of the article is positive, negative or neutral to Climate Change:

1   Positive
2   Negative
3   Neutral

Examples of negative mentions

An article that is clearly critical of Climate Change ("State defers plan to oppose UNESCO status," *TOI*, 7 July 2012).

An article that blames politicians or associates them in a negative way with Climate Change ("Kasturirangan panel report: politicians involved in mudslinging," DH, 24 March 2017).

Examples of positive mentions

Generally speaking, any article that shows a solution-oriented approach in a positive tone (featuring people involved in conservation efforts, tiger census showing increase, etc.)

Column L: type of Climate Change

Here, we distinguish between the types of Climate Change focused in the story:

1   Threat to Western Ghats
2   Threat to community

3   Threat to economy
4   All of the aforementioned

Column M: Western Ghats

Here, we distinguish between the types of coverage to Western Ghats focused in the story:

1   Related to ESA reports
2   Related to conservation/wildlife
3   Related to tourism/development
4   Related to sustainability
5   Related to forest community
6   Related to energy/resources/hydel projects

Column N: forest communities

Is the article trying to raise awareness among forest communities about Climate Change?

1   Yes
2   No

Column O: policy-related

Here, we should distinguish between the types of coverage of policy-related issues focused in the story:

1   Knowledge about policy in existence
2   Attitude of forest communities toward policy
3   Implementation of policy
4   Not related to policy

Column P: communication of Climate Change

Here, we should identify how the article is trying to communicate about environment issues:

1   Convincing the public to take steps to mitigate Climate Change
2   Featuring crusaders/everyday heroes of sustainability
3   Context of Paris Agreement/UNFCCC
4   Fear psychosis

Column Q: challenges for media to communicate Climate Change

Here, we identify the challenges faced by journalists/media houses to communicate issues:

1   Persuade readers toward immediate behavioral change
2   Cultural barriers and social norms
3   Community leaders, celebrities, governments as communicators

Appendix B

(a) Constitution of expert committee to assess the carrying capacity of Western Ghats region in Karnataka

(i) Order, (ii) Amendment to the Order

Minutes of the Meeting of MPs of Western Ghats region held on 11.08.2016
HLWG-Report-Part-1_0
HLWG-Report-Part-2
Directions dated 13.11.2013 under Section 5 of EP Act, 1986 on Western Ghats ESA
First Draft Notification dated 10.03.2014 on Western Ghats ESA
Second Draft Notification dated 04.09.2015 on ESA for Western Ghats
NGT order dated 15.2.2016 in M.A in W.P. 202 of 1995
S.O. 667 (E) [28.02.2017] Draft Notification declaring Eco Sensitive Area around Western Ghats, India
List of Hon'ble Members of Parliament Participated the Meeting of Hon'ble Minister (EF & CC) with the Hon'ble Members of Parliament of Western Ghats Region held on 11th August, 2016
S N Name House (Rajya Sabha/Lok Sabha)

1   Shri Sharad Pawar RS (Maharashtra)
2   Shri Joice George LS (Kerala) (Idukki)
3   Shri K Soma Prasad RS (Kerala)
4   Shri Joy Abraham RS (Kerala)
5   Shri C P Narayanan RS (Kerala)
6   Shri Vinayak Bhaurao Raut LS (Ratnagiri-Sindudurg) Maharashtra)

7 Smt. P.K Shrimati Teacher LS (Kerala)
8 Shri Nalin Kumar Kateel LS (Karnataka) (Dakshina Kan-nada)
9 Shri Anant Kumar Hedge LS (Karnataka) (Uttara Kan-nada)
10 Dr. Shashi Tharoor LS (Kerala) (Thiruvananthapuram)
11 Dr. A Sampath LS (Kerala) (Attingal)
12 Shri Anto Antony LS (Kerala) (Pathanamthitta)
13 Shri Jose K Mani LS (Kerala) (Kottayam)
14 Shri Dilip Gandhi LS (Maharashtra)(Ahmednagar)
15 Shri Tiruchi Siva RS (Tamil Nadu)
16 Shri Oscar Fernandes RS (Karnataka)
(Accessed from website of Ministry of Environment, Forests and Climate Change)

(b)

GOVERNMENT OF INDIA
MINISTRY OF ENVIRONMENT, FOREST AND CLIMATE CHANGE

RAJYA SABHA
UNSTARRED QUESTION NO. 30
TO BE ANSWERED ON 25.04.2016

Eco-sensitive areas in Western Ghats

30. SHRI HUSAIN DALWAI:

Will the Minister of ENVIRONMENT, FORESTS AND CLIMATE CHANGE be pleased to state:

(a) what is the progress on declaring eco-sensitive areas (ESAs) in Western Ghats;
(b) how many States have submitted their report on demarcating ESAs by physical verification;
(c) what is the procedure to be followed by the State to hold public consultations before demarcating ESAs by physical verification;
(d) how many public consultations were held in Maharashtra for demarcation of ESAs, village-wise; and
(e) whether report of the State would be accepted, even if no public consultation is held and if so, the reasons therefor?

ANSWER

MINISTER OF STATE (INDEPENDENT CHARGE) FOR ENVIRON-
MENT, FOREST AND CLIMATE CHANGE
(SHRI PRAKASH JAVADEKAR)

(a): The Ministry of Environment, Forest and Climate Change has brought out a fresh notification dated 4th September, 2015 for declaring ecologically sensitive area in the Western Ghats in supersession of the earlier draft notification issued on 10th March, 2014 incorporating provisions to clarify and to dispel the apprehensions and concerns raised by various stakeholders.

(b): Reports of the States of Kerala, Goa, Karnataka, Maharashtra and Gujarat. Proposals have been received for demarcating ecologically sensitive areas (ESAs) in their respective States.

(c) to (e): The Ministry had given an opportunity to State Governments of the Western Ghats region for undertaking the exercise of demarcation of eco-sensitive areas by physical verification. No specific procedure was prescribed by the Ministry for holding public consultations before demarcating ESAs by physical verification.

The Government of Maharashtra has undertaken public consultations in 2,025 villages identified as being part of Western Ghats Eco-sensitive Areas as part of physical demarcation.

Acceptance of reports of the State Governments, submitted after undertaking demarcation of ESA by physical verification in Western Ghats, is determined mainly by identification ecologically sensitive areas on scientific basis. However, the States were advised to also resolve the apprehensions/concerns expressed by various stakeholders of Western Ghats from time to time.

Note

1 This could include a news release from this organization.

INDEX

Note: **Bold** page numbers denote references to illustrations.

Printed in the United States
by Baker & Taylor Publisher Services

Printed in the United States
by Baker & Taylor Publisher Services